D0903366

ARTIFICIAL SUBSTRATES

ARTIFICIAL SUBSTRATES

JOHN CAIRNS, JR.
Editor

230 Collingwood, P.O. Box 1425, Ann Arbor, Michigan 48106

Library of Congress Catalog Card Number 81-69073
ISBN 0-250-40404-4

Manufactured in the United States of America

Butterworths, Ltd., Borough Green, Sevenoaks,
Kent TN15 8PH, England

FOREWORD

Aquatic organisms have used artificial substrates as habitats since the first caveman threw an implement or potsherd into a body of water. It was not until the present century, however, that artificial substrates have been carefully evaluated as the means for assaying the characteristics of a natural body of water. At first they were simply used as collectors of a given group of organisms, with no effort made to establish whether they were constructed to collect similar samples from a given area, or from different areas. Nor was any effort made to determine if the species that colonized and grew on these substrates were representative of the species living in an area.

Within the last 30 years, many efforts have been made to answer these questions. Some of the notable among these are John Cairns, Jr. and his associates in the study of Protozoa; William T. Mason and his associates in the studies of baskets and multiplate macroinvertebrate samplers; and Ruth Patrick and her associates in their studies of diatoms.

Many articles have been written; some concluded that the use of artificial substrates was a reliable method of estimating the species present in an area, and others concluded that they were not. To have artificial substrates function in a predictable way requires that careful consideration be given to how the samplers are constructed, how they are placed in the environment, and how the collections so obtained are analyzed. For example, the material from which samplers are made must be tested to see if it offers optimum sites for colonization and growth. Samplers must be placed in light and current conditions that are most favorable for colonization and growth of the specimens reaching the samplers. They must be placed so that they are not subject to catastrophic events from boat and barge traffic. The samplers must be placed in similar habitats if they are to be compared. Samples analyzed must be large enough to be truly representative. For example, in studying microorganisms, specimens are counted. Often, 500 or 1000 specimens are not enough if dominant species are present and one wishes to determine the species in the sample that are characteristic of the natural environment.

Artificial substrates are the means of making many different types of studies concerning the aquatic environment. They offer (1) a means of studying colonization rates, (2) growth and reproduction rates, and (3) if large enough, they furnish sufficient specimens to model the structure of the community in question. Knowing the time of exposure also excludes an important variable in interpreting results. The information derived from the use of artificial substrates is closely correlated with the time and ingenuity involved in designing, in determining placement, and in analyzing the sample collected.

Ruth Patrick
The Academy of Natural Sciences
Philadelphia, PA

Dr. John Cairns, Jr. is University Distinguished Professor in the Biology Department, and Director of the University Center for Environmental Studies at Virginia Polytechnic Institute and State University, Blacksburg. He received his PhD in Zoology and his MS in Protozoology from the University of Pennsylvania, his AB in Biology from Swarthmore College, and completed a postdoctoral course in isotope methodology at Hahnemann Medical College, Philadelphia. He was Curator of Limnology at the Academy of Natural Sciences of Philadelphia for 18 years, and has taught at various universities and biological laboratories.

He received the Charles B. Dudley Award in 1978 for excellence in publications from the American Society for Testing and Materials, and the Presidential Commendation in 1971. A member of many professional societies, he is Chairman, Committee on Ecotoxicology, of the National Research Council. Dr. Cairns has been consultant and researcher for the government and private industries, and has served on numerous scientific committees. His more than 700 publications encompass 24 books, including a 1980 Ann Arbor Science publication, *The Recovery Process in Damaged Ecosystems,* monographs, numerous chapters in books edited by others, scientific papers, abstracts, book reviews and congressional testimony.

Artificial substrate: An artificial substrate is a device placed in an aquatic ecosystem to study colonization by indigenous organisms. Although the device may be unnatural in composition, location, or both, most of the biological processes that occur on it appear to be quite similar to those occurring on natural substrates.

PREFACE

The use of the word "artificial" in referring to introduced substrates in aquatic ecosystems has had some unfortunate consequences. The most serious is that many research investigators believe that the species that colonize these substrates are not characteristic of those that colonize natural substrates, and that the community structure and functional dynamics are quite unlike those that occur on natural substrates. Although the introduced substrates may have a chemical composition alien to natural environments (e.g., polyurethane), they are essentially inert, a characteristic common to many other natural substrates, such as rocks. The fact that they are free of aquatic organisms when introduced makes them no more alien to the ecosystem than a rock that has tumbled into a stream in a rockslide. Their location may be atypical (e.g., a substrate positioned in the epilimnion of a lake), but the species that have colonized the substrate after a week or so of exposure are those that inhabit the surface of solid substrates in the benthic area of the lake or in the shallow water along the shoreline. Most introduced substrates are chosen because they do not decompose and are not quickly eroded by various water actions. In short, they are quite inert and merely furnish a surface for attachment of both living and nonliving materials. Thus, although their composition may be artificial, the biological events that occur on them appear to be quite similar to those that occur on natural substrates of similar structural characteristics.

Even if one accepts the fact that events on introduced substrates are quite similar to those on substrates already present in the aquatic ecosystem, the question is, "why study introduced substrates when there are so many natural substrates available for obtaining collections?" A few of the major reasons follow:

1. The age of the community on natural substrates is not known, but this can easily be determined when introduced substrates are used.

2. Two microhabitats of precisely the same structure, composition, surface area, and so on are exceedingly rare to find. As a consequence, when differences are found in the biota inhabiting two dissimilar microhabitats, the differences cannot be pinpointed with certainty to the habitat itself or to some other characteristic, such as water quality differences. Lack of uniformity is not a problem with introduced substrates.
3. Natural substrates can often be found that are quite similar, but they are in the wrong location. Introduced substrates may be positioned anywhere so that the exposure to current, amount of sunlight and other important characteristics is as similar as possible in natural ecosystems.
4. Statistical reliability is more easily achieved when a number of substrates of identical composition, structure, position, etc., can be placed in ecosystems to assure that an adequate sample size for statistical evaluation is possible from a particular type of microhabitat. Even in the unlikely event that a large amount of uniform habitat is available for sampling in comparable situations, the problem of determining the age of the community on each of these natural substrates still exists. One can assume that all of the communities are of the same age but, with introduced substrates, one knows that this is the case.
5. Introduced substrates provide the opportunity to study succession and the colonization process itself at times determined by the investigator rather than having to wait for a natural event to make this possible.
6. There is compelling evidence that the early stages of colonization furnish useful information that is qualitatively and quantitatively different from the information derived from mature communities.

The chapters in this book discuss both marine and freshwater systems and both microbial and macroinvertebrate species. The book was designed to illustrate the opportunities for using artificial substrates in both theoretical and applied research and does not attempt to cover all possible aspects.

John Cairns, Jr.

ACKNOWLEDGEMENTS

This book is a result of a symposium sponsored by the American Microscopical Society. The editor gratefully acknowledges the society's continued support.

The editor gratefully acknowledges the following individuals who spent time and effort in the review of the papers that are incorporated in this volume.

Stuart S. Bamforth
Newcomb College
Tulane University

E. Fred Benfield
Virginia Polytechnic Institute
and State University

John O. Corliss
University of Maryland

Paul K. Dayton
University of California

Conrad Istock
University of Rochester

Robert J. Livingston
Florida State University

Ralph Mitchell
Harvard University

George W. Salt
University of California

Daniel Simberloff
Florida State University

Eugene Small
University of Maryland

R. Jan Stevenson
University of Michigan

John P. Sutherland
Duke University
Marine Laboratory

James Ward
Colorado State University

Jackson R. Webster
Virginia Polytechnic Institute
and State University

ROYALTIES

Some of the contributors to this volume wish to keep in the spirit of the previous volume that I edited for *Ann Arbor Science*, entitled *The Recovery Process in Damaged Ecosystems*, for which all royalties were given to the MacArthur Fund of the Ecological Society of America. Royalties from *Artificial Substrates* will be donated to the Spencer-Tolles Fund of the American Microscopical Society by the following contributors: S. S. Bamforth, J. Cairns, Jr., D. V. Ellis, L. G. Harris, M. S. Henebry, K. P. Irons and D. M. Rosenberg. R. M Osman and A. Schoener are donating all royalties from this volume to the MacArthur Fund of the Ecological Society of America. J. F. Flannagan is donating his royalties to the International Permanent Committee for Ephemeroptera Conferences Scholarship Fund and V. H. Resh is donating his royalties to the North American Benthological Society.

CONTENTS

CHAPTER 1

ARTIFICIAL SUBSTRATES IN MARINE ENVIRONMENTS

Amy Schoener

Department of Oceanography
University of Washington
Seattle, Washington 98195

Artificial substrates serve a useful function in investigating a wide variety of problems. Here they are employed to compare colonization events at two localities in Puget Sound, Washington, where colonization patterns of epibenthic macroorganisms are examined. The dominant early colonists at these sites belong to different phyla: one area is primarily covered by tunicates, the other by bryozoans. Though phyletic differences exist among colonists, colonization curves show marked similarities. So, too, do the functional groups of space occupiers present; initial dominance by colonial organisms changes to dominance by solitary species as time progresses. This trend is a striking contrast to that reported for tropical localities. Although qualitative differences exist between communities on identical artificial substrates, rates of colonization, panel coverage and numbers of species present are very similar.

INTRODUCTION

Only a fraction of the manmade products jettisoned into the oceans are intentionally slated for later removal. Admirably, some artificial substrates, categorically referred to as fouling panels, are so designated from the first; they constitute the subject of this chapter. Typically, an array of identical panels is submerged simultaneously and its position marked. Later, individual panels are returned to the surface at intervals until all have been retrieved. The sequence of events on individual panels in a series is equated with the time course of events on an average panel. If continuity of technique marks any indication of its success, then surely this must be considered successful. Published reports utilizing destructive sampling span several decades [1-4]. More recently, modifications of this procedure have allowed individual panels to be followed through time by nondestructively observing them in seawater, then repositioning them to allow further colonization [5-10]. This alternative approach is now widely adopted in field research, including that reported here.

In comparison to their size, fouling panels have contributed disproportionately in matters of military and scientific import. Generally of small dimensions and, therefore, quite manageable, panel arrays have been deployed both nearshore and offshore to establish type and biomass of fouling growth worldwide [11-16]. Primarily positioned in shallow water, their submergence in deep water has led to surprising discoveries, among them the first evidence of rapid colonization processes in the deep sea [17], a contrast to the slow recolonization of soft sediments there [18].

Recognizing that the investigation of general patterns in ecological research often profitably crosses taxonomic bounds, this chapter attempts to document some quantitative patterns in marine biogeography considering a variety of colonists on artificial substrates positioned at two Puget Sound localities.

A PLACE FOR ARTIFICIALITY IN NATURE

Although several compelling reasons favor the choice of artificial substrates over their natural counterparts, these would be less relevant were not similar taxa and processes occurring on both. Fortunately, these questions have been answered affirmatively, based on the following observations. First, similar colonists appear on natural and artificial substrates; this is true of plastic and real sponges [19]; rocks and those concrete blocks or slate panels designed to mimic them [7, 20, 21]; and, to a lesser extent, for mangrove shoots and asbestos panels [22]. Second, similar processes occur on both artificial and natural hard substrates in marine environments when the two are compared [20]. Third, certain processes tend to be relatively unimportant at the same locality on both artificial and natural substrates. For example, Jackson [2] points out that predation is barely noticeable in either natural cryptic environments or on artificial substrates in cryptic environments when these are observed.

If nature is to be mimicked, choice of substrate type must be considered carefully; some artificial substrates can be selected that closely duplicate the surface texture and mineralogy of the objects they are designed to mimic [6,7]. This might be of greater concern were the fouling organisms not as widely tolerant of differing surfaces as they appear to be, e.g., sea snakes are colonized by as many phyla of fouling organisms as are found on artificial substrates [23], and living mollusk shells accumulate considerable diversity of growth as well [24].

Choice of artificial rather than natural substrates often is made primarily for utilitarian reasons. Good experimental protocol depends on careful experimental design and sufficient replication. The major points favoring use of artificial substrates relate to these concerns, particularly to the degree of control the experimenter retains, substrate size and type, extent of replication, even ability to minimize certain types of disturbance, all can be altered. In preference to seeking natural substrates of a specific size from the suite of those available, it is more convenient to create a series of uniform substrates using artificial materials whose dimensions can be controlled more accurately. Additionally, since artificial substrates can be standardized, direct comparisons between different localities can be made without introducing further components of variability associated with substrate type. Lastly, sufficient replication can be attained readily if artificial substrates are utilized.

COLONIZATION EVENTS ON ARTIFICIAL SUBSTRATES

In most instances, monitoring colonization of artificial substrates is not directly comparable between localities because individual investigators vary substrate size, materials and observation techniques. Rarely can colonization patterns on identical substrates be compared among sites from different latitudes. In the few instances where this is possible, comparisons indicate that the initial rates of increase in the number of sessile species accumulating with time are inversely related to latitude [25]. The data that delineate these trends are derived from several sites in the northern hemisphere spanning tropics to arctic.

These data do suffer several deficiencies, however, particularly concerning replication, distance between test regions and duration of the experiment. Since one temperate locality included in that comparison was Puget Sound, Washington, an opportunity presented itself to compare colonization in greater detail. Of particular concern are the rates of accumulation of species and percent cover on panels, although attention is also devoted to analysis of the types of colonists involved. The patterns for different sites at the same latitude (ca. 48°N) are investigated to determine the generalities and inherent variability.

Experimental Procedure

Replicate textured white Formica® squares (412.9 cm^2) were each fitted and glued to supporting Formica surfaces of equal dimensions. These substrates then were submerged from floats and positioned horizontally about one meter below the water's surface. Only the undersurface of these substrates is observed, as this side accumulates the least sediment and the most diverse biotas [10]. Each panel was suspended independently by a rope passing through its center and attached to a weight one meter beneath. Therefore, it could not be overturned unintentionally. Since weights remained suspended above the bottom, benthic organisms could not reach these substrates directly. Edges of adjacent substrates were separated from one another by at least 36 cm to minimize lateral interactions. Each panel was observed nondestructively in seawater for less than one hour's duration, then repositioned for further colonization. During the early stages of panel development, these monitorings were more frequent than at later stages of the approximately two-year observation period.

At each observation time two measurements were recorded. The first was obtained by carefully noting all species greater than 0.5 mm that were attached directly to the panel surface; this allows an estimate of total species richness of panels. The second measure yields an estimate of the percent cover of panels. It is based on point sampling. Random points plotted on a transparent sheet are placed atop panels, and the number of points hitting attached species are tallied. This provides an estimate of percent cover with 5% accuracy [8].

For comparative purposes, dates of initiation of the experiments were synchronized between localities, although studies took place in successive years. Seasonal changes in colonization rates at temperate localities are well documented [7] and indicate that submergence time is critical to the shape of the colonization curve. Submerging both sets of panels in midspring minimized this seasonal variation but did not eliminate the possibility that annual settlement differences might arise [6,8]. This may account for some of the differences in species composition observed between localities, although other factors are also considered. The aim of this analysis is to investigate the generality of patterns existing in the presence of taxonomic variations; differences in species composition are welcomed.

Replicability Among Identical Substrates

No doubt, any determination of pattern is dependent on the accuracy and precision of its portrayal; this itself is a function of sufficient replication to establish its validity. Determination of what constitutes sufficient sample size rests both on the variability inherent in the parameters examined and the types of questions asked [26]. Despite the extensive literature on fouling, studies on the degree of variability in total species counts or percent cover of individual substrates has, until recently, been a neglected area of research. Several short-term surveys report minimal variability between replicate panels [27,28]. Osman's [7] observations over a substantial time period also indicate that variability within a series is relatively low, as do some data collected in Hawaii [25]. These results agree with findings for substrates submerged in freshwater, where only a few replicates are sufficient to portray accurately species diversity [29, 30].

On the other hand, if investigators concentrate on the identities of individual colonists, a greater variability per series is reported [5,8]. These studies stress the individual histories and species compositions of panels. In part, the discrepancies between their results and the other findings are related to just this emphasis, with less variability appearing among species totals than among species identities. This suggests that a certain degree of species substitution is possible. Osman [7] considers that replacement of species by others may not substantially affect the trophic dynamics of his artificial plate system, since all sessile colonists there are filter-feeders.

To identify the extent of variability among artificial substrates in Puget Sound, an extensive series of identical substrates was submerged from floats in Phinney Bay (Figure 1) and analyzed periodically [31]. By observing 26 of 150 identical panels at each examination period, mean total number of sessile species and percent covers were computed. Testing the data for deviations from normality by the Kolmogorov-Smirnoff test [32] indicated that assumptions of normality were not violated ($P<0.05$), and these data were not transformed. The 95% confidence limits were then computed for hypothetical subsamples of decreasing size in each series. The resulting plots indicate their decreasing width with increasing sample size, giving them a trumpet shape [33]. An example of this variability for 1½-year-old panels is given in Figure 2. The flare in the trumpet decreases substantially after samples of between

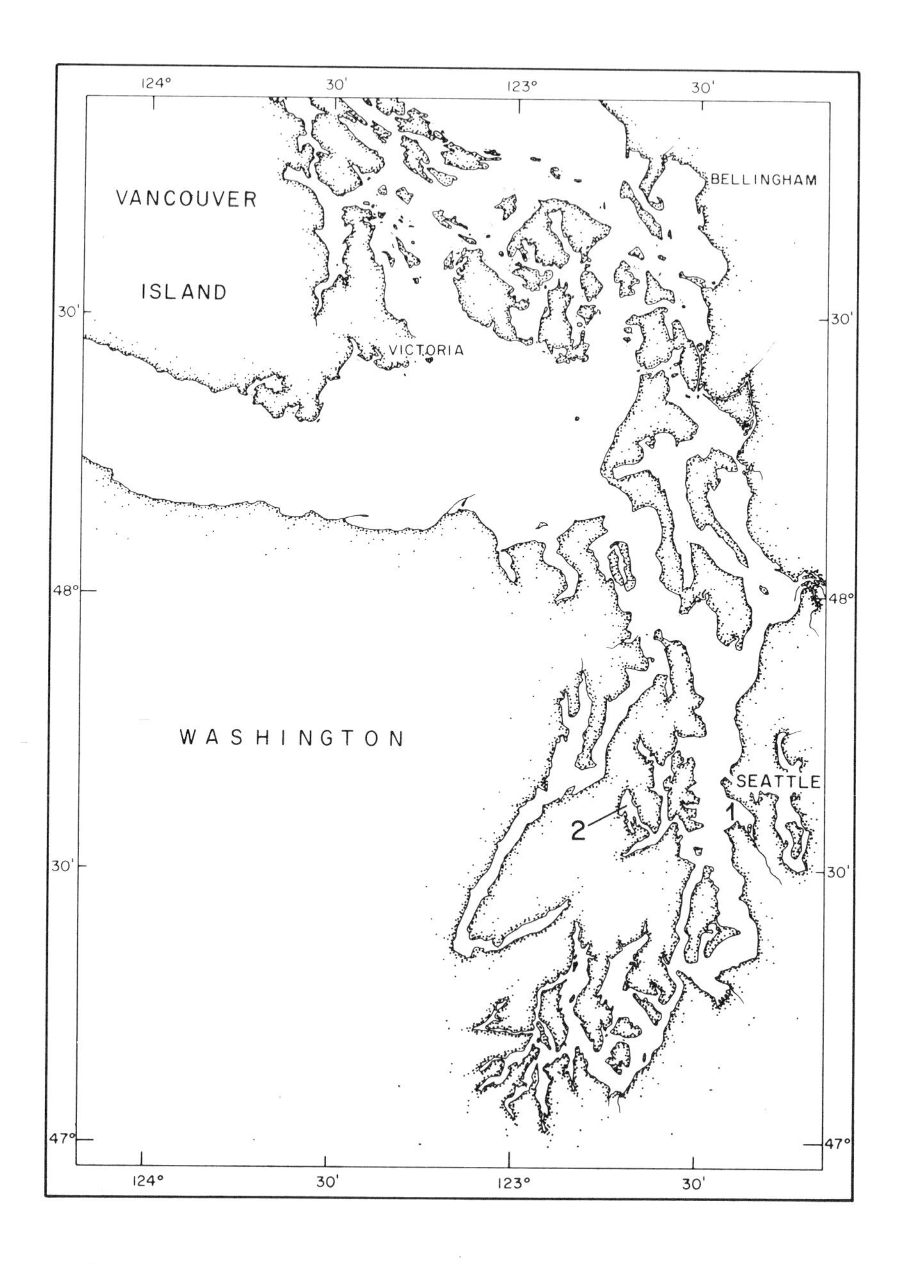

Figure 1. Map of Puget Sound localities. 1 = Elliott Bay site; 2 = Phinney Bay site.

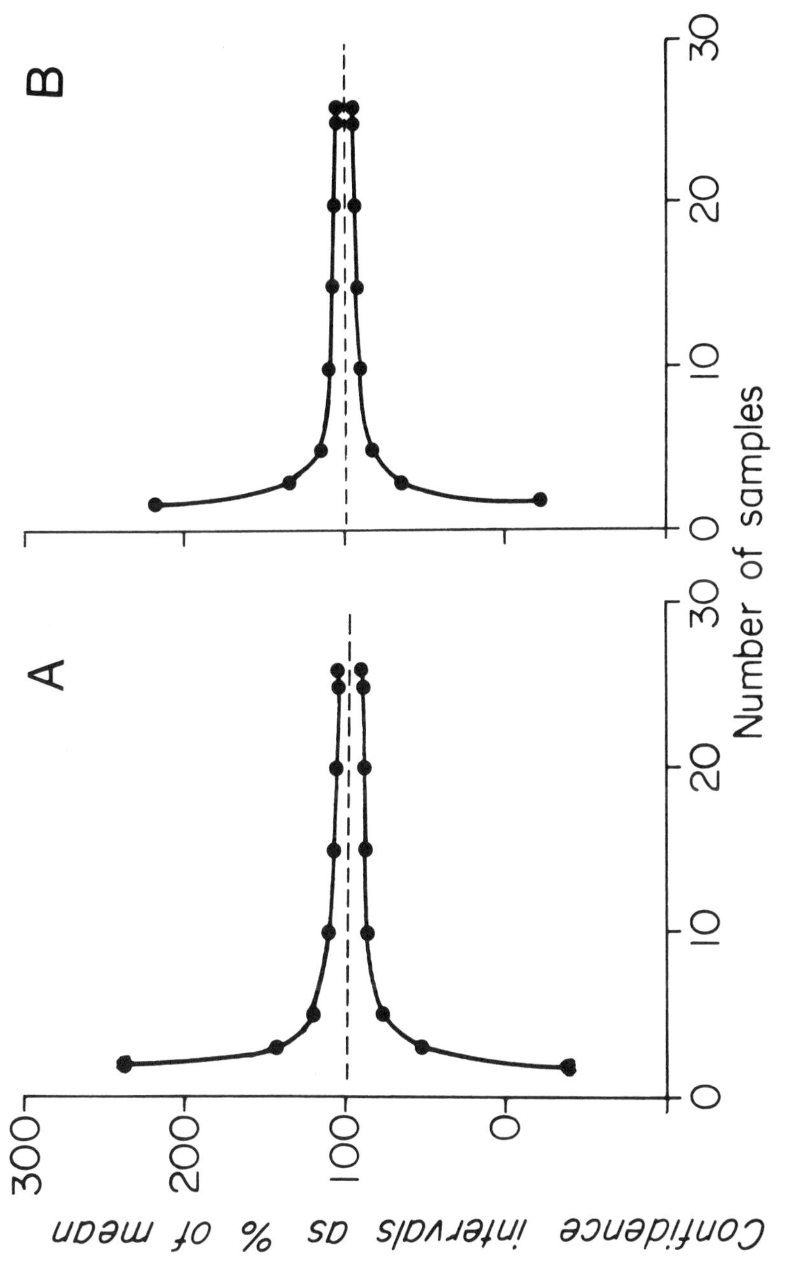

Figure 2. The upper and lower 95% confidence limits (expressed as a percentage of the mean) plotted versus sample size (number of panels). The data represent mean number of sessile species (A) and mean percent cover (B) for 26 panels after 1½ years of colonization at Site 2.

6 and 10 are observed. This suggests that if confidence limits within 10% of the mean are acceptable, small sample sizes are adequate to portray accurately species numbers and percent cover in Puget Sound. While increased sample size often results in additional information, observing greater numbers of samples does not substantially reduce confidence limits around the parameters in question. This is true both for number of attached species (Figure 2A) and for percent cover (Figure 2B). These results were typical of those obtained throughout the first year of study [31]; however, somewhat wider confidence intervals were computed for percent cover estimates during the first observation period when total coverage was scant. On the basis of these trumpet diagrams, if the parameters of species numbers or percent cover are of interest, then observation of 6 to 10 panels would seem to be a reasonable sample size, since the benefits of larger sample sizes diminish quite rapidly beyond this.

Differences in abundance of particular species may offer some explanation of the variability observed in total species counts between otherwise identical sets of panels positioned equivalently. At Elliott Bay (Figure 1), for instance, species numbers on 13 identical smaller panels (6.4 cm^2) were compared over one year's time. The deviation of each panel from the median number of species present for the series gives an idea of this variability (Table I). It is evident that although many panels show some deviation in one direction or another, in 12 of 13 panels this is only slight. In one case (P-5), however, this deviation is substantial and positive. Over the dates considered this panel almost always averages a greater number of species than do the remaining panels. Comparisons of species lists for all panels are similar during initial colonization. On these panels an encrusting bryozoan *Hippothoa* and a hydroid *Clytia* account for the majority of percent cover. All panels except the deviant one had extensive coverage by the bryozoan at some time during early colonization (100% in nine cases and at least 92% in three additional cases). The sole panel on which *Hippothoa* coverage never exceeded 70% was the P-5 panel, suggesting that if the encrusting bryozoan did cover panels extensively during early colonization, species richness in subsequent periods was lower than if this did not occur. The ability of one encrusting bryozoan to influence species' presence is shown nicely by Sutherland's [34] experimental removal of a similar species; this indicated that the encrusting bryozoan was inhibiting settlement by various tunicate, sponge and barnacle species.

Although compositional differences exist between individual panels and may well be the result of presence or absence of particular species, quantitative descriptions of the colonization process can accommodate these variations while still portraying many of the general aspects of colonization.

Table I. Deviation of Individual Panel's Species Numbers from Median Species Number During Later Experimental Period (6.4-cm^2 panels)

	Week												
Panel	78	67	58	49	38	33	29	25	21	16	13	11	Total Deviation
P-10	−3	−1	+2	−2	+4	+1	+1	0	0	0	−1	0	+1
P-7	+1	0	−1	0	−1	−1	0	0	+1	−1	0	−1	−2
P-1	0	−2	−1	0	−1	0	−1	0	+2	−2	0	0	−5
P-2	0	0	0	+3	−1	0	0	−1	0	−3	0	−1	−3
P-3	0	−2	−1	−2	0	+2	−1	−1	+1	0	−2	−2	−8
P-4	0	+2	−1	+3	−2	0	−1	−1	−1	−2	−1	−1	−6
P-5	+1	0	+1	+4	+4	+5	+5	+3	+4	+5	+3	+4	+39
P-6	−1	0	+1	+4	+1	0	0	+1	0	0	0	0	+6
P-8	0	+1	+1	0	0	+1	0	0	+1	−2	+1	−4	−1
P-9	−2	−2	+1	+2	−1	0	−1	+1	+2	0	0	0	0
P-11	−2	−1	+1	+4	+1	0	−1	−3	−1	+1	−1	0	−3
P-12	−1	0	−1	0	+3	0	−1	0	0	+1	−2	−2	−3
P-13	0	+1	−2	0	−1	−1	0	−1	0	+2	0	+1	−1
Median	5	5	6	6	6	5	5	6	5	5	5	5	

Study Sites

In both cases, experiments were conducted from floating docks in the central basin of Puget Sound, Washington, and identical materials and dimensions were employed at the two localities described below (Figure 1).

Site 1—Elliott Bay: Located off Seattle proper, this large bay is a commercially active harbor. Panels were suspended from docks at the public Seacrest Marina. The experimental site was relatively free of both pollutants and silt carried by river runoff emptying into the bay, since flow patterns in the bay carry riverborne materials toward the shore opposite the marina. Six panels were submerged in mid-June 1977; water depth below them was about 3 meters at lowest low tides.

Site 2—Phinney Bay: Located west of Seattle, Phinney Bay is a small bay off the main channel of Puget Sound, near Bremerton, Washington. As such, it is fairly protected; floats and artificial substrates here survived unscathed the winter storm that destroyed the nearby Hood Canal bridge. This quiet bay experiences no commercial traffic, although a private yacht club (Bremerton Yacht Club) provided the floats from which 150 panels were submerged simultaneously in mid-June, 1978. Water depth below panels was about 4 meters at lowest low tides.

The Colonists

While algae are peripherally important, invertebrates and tunicates constitute the majority of attached colonists. At each locality a total of about 60–70 species of tunicates, sponges, coelenterates, algae, mollusks, polychaetes, bryozoans and arthropods have been identified as sessile. To reach these isolated substrates, individuals must traverse the water column from other hard substrates. Note here the distinct parallel between these isolates and terrestrial islands, whose colonists likewise have to cross an alien medium to reach appropriate new areas [7].

For marine colonists, crossing a water gap presents problems of varying degree. First, their ability to immigrate depends on having a pelagic stage in their otherwise sedentary life cycle. Second, while pelagic, they are exposed to both predation and the vicissitudes of settlement; and, if they are not eaten, currents may remove them from suitable localities. Whereas some "sessile" marine species may detach as adults and colonize new substrates (e.g., anemones), in actuality this rarely occurs in the marine realm. Osman [7] estimates that at least 90% of the colonists reaching artificial substrates must immigrate as larvae, and that larval colonization was largely responsible for their presence even among species whose adults are capable of mobility. Immigration, therefore, occurs during the early pelagic phase in the life cycle of species characterized by sessile adulthood.

Colonization Sequence

Hydroids are the first noticeable macroscopic growth occurring at either Puget Sound locality, although the species involved vary. Figure 3 indicates the total percent cover of panels by hydroid growth. Subsequently, either extensive bryozoan coverage accumulates in Elliott Bay (Figure 4) or extensive tunicate cover develops at Phinney Bay (Figure 5). Several species of each of these taxa comprise the total percentages summarized in the figures, although Phinney Bay has a preponderance of one solitary tunicate species (*Corella willmeriana*) early in colonization.

Colonization Patterns

There is some overlap in species composition between sites, yet taxonomic differences prevail and prevent many direct comparisons. By momentarily divesting these data of such considerations, one may compare colonization rates at these localities. The total number of species present is compiled and plotted against time, yielding colonization curves [35].

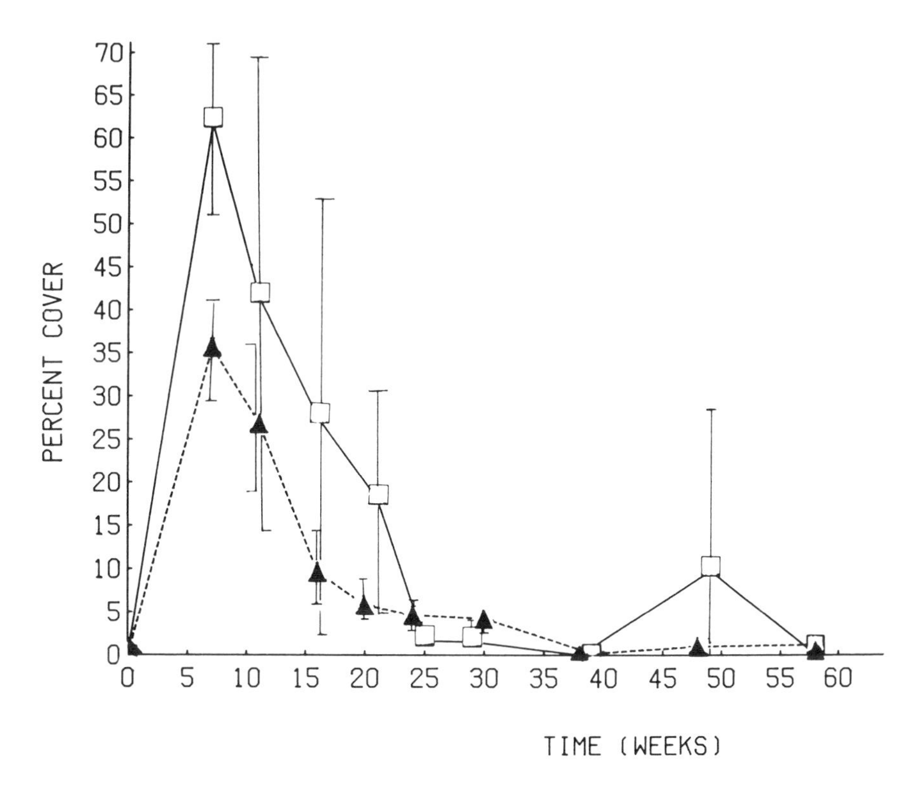

Figure 3. Total mean coverage of panels by all hydroids, plotted against time. Squares represent panels at Site 1 (Elliott Bay) and triangles represent panels at Site 2 (Phinney Bay). The 95% confidence intervals are given as bars.

These colonization curves are shown in Figure 6 and are directly comparable since substrate type, dimensions and observer are identical. Curves rise rapidly, as would be expected in late spring, a time of increased immigration onto panels in temperate areas over that occurring during winter [7]. They level off after several months without experiencing any pronounced overshoot in species numbers [36]. Both curves can be broken after the fourth month and simple linear regression run on the early portions of each curve; this shows that both curves have significantly positive slopes ($P < 0.05$), which are statistically indistinguishable from one another (1.39–1.45 species/week, $P < 0.05$). After several months these curves reach similar species numbers (ca. 20–21). If the curves are examined during this asymptotic phase between weeks 17 and 49, the slopes of the curves are slight (0.04–0.08 species/week) and neither the slopes nor the intercepts are significantly different from each other

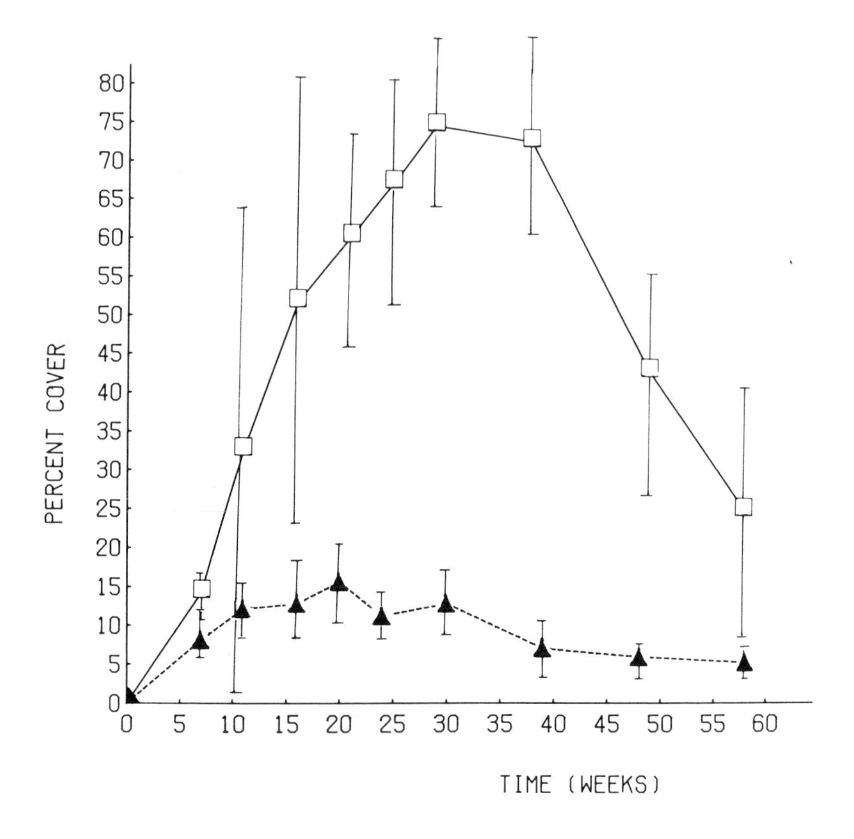

Figure 4. Total mean panel coverage by all bryozoan species, plotted against time. Symbols as above.

(P<0.05). The later portions of these curves indicate a decline in species numbers and, in one case, a subsequent rise, perhaps the beginning of an oscillatory phase [7].

In late spring, the rate at which artificial substrates are covered by species is rapid at both localities. Figure 7 indicates that at Phinney Bay percent cover rises quickly and soon oscillates around 75% after three months have elapsed. Even higher coverage subsequently develops toward the end of the first year, although the maximum values obtained earlier are maintained throughout most of that year. After this time, coverage oscillates around 80% for the experiment's duration. At Elliott Bay, initial coverage of panels is similarly rapid, reaching 90% during the same three-month period of initial colonization. Values close to this are maintained throughout the remainder of the experiment. Both locali-

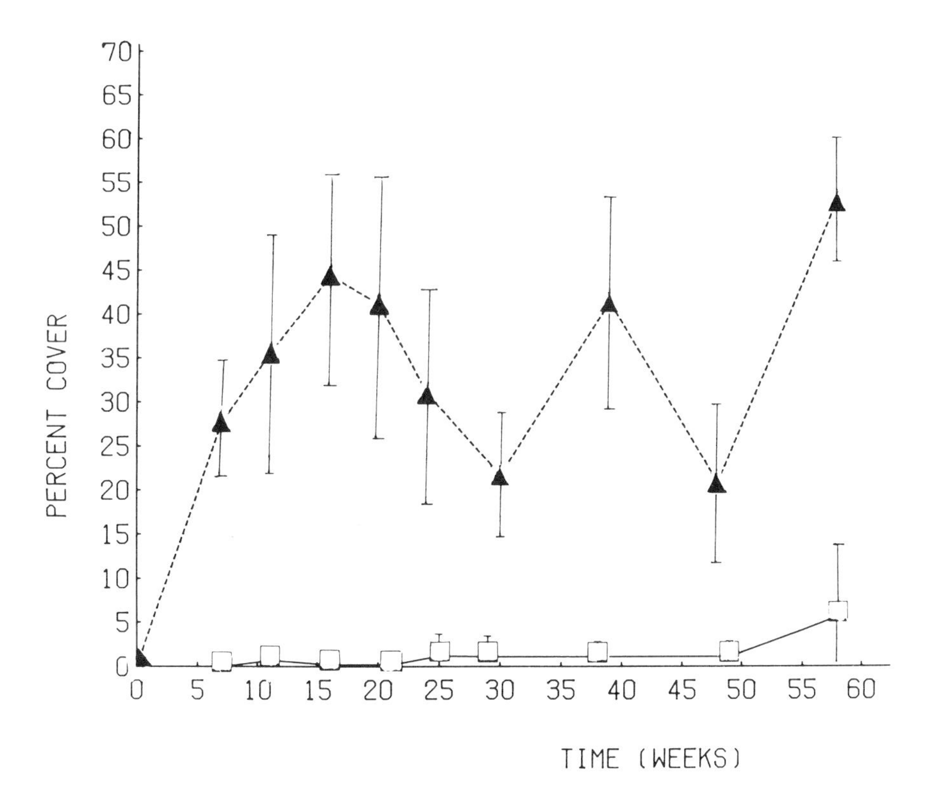

Figure 5. Total mean panel coverage by all tunicate species, plotted against time. Symbols as above.

ties experience fairly stable species numbers during much of the first year, a phenomenon beginning close to the time at which panel coverage begins to stabilize at its high values. This suggests that decreasing availability of space influences the rate at which new species can be added to panels. Since species can affect a somewhat greater area than they actually occupy, coverages of less than 100% can be maximally effective.

The initially high percent cover observed is maintained even though species numbers begin to decline at Elliott Bay, suggesting that competitive interactions are less responsible for this decline than are life history factors. Osman [7] proposes several variations in immigration and extinction rate curves, which can explain dips and rises in the colonization curve. His Case C, in which extinction rates vary seasonally along with immigration rates, conforms most closely to the shape of the

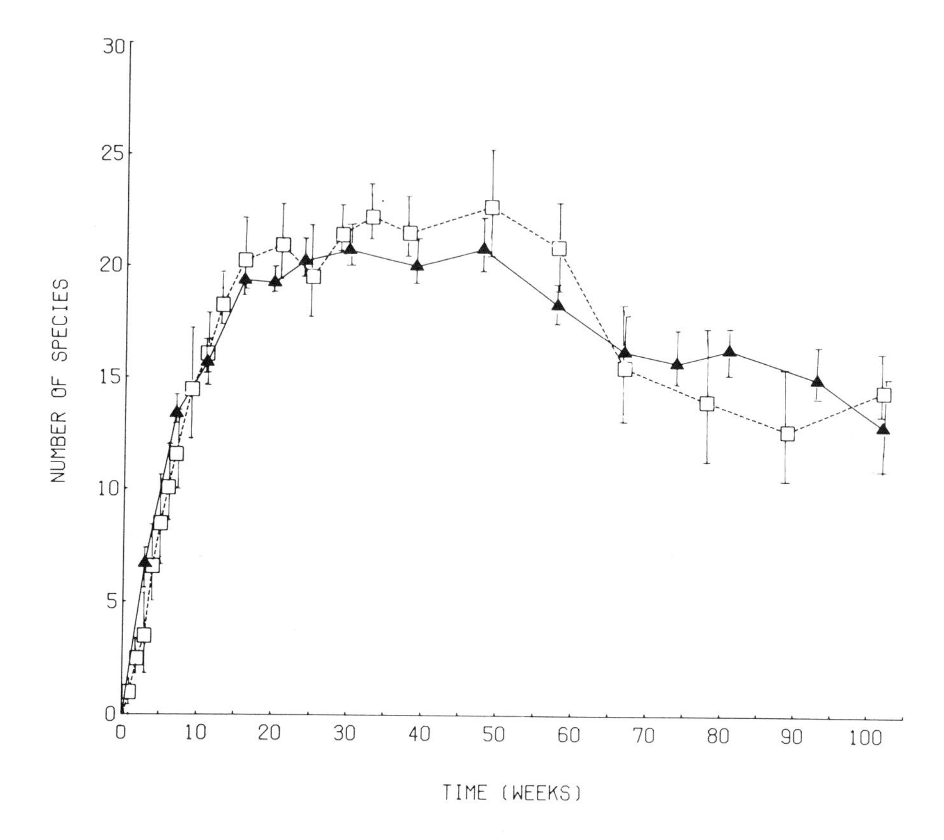

Figure 6. Colonization curves plotting mean number of sessile species at Sites 1 and 2 over time. Symbols as above.

observed Puget Sound colonization curves. His model predicts decreases in species numbers earlier than those actually occurring at these localities. Subsequent increases, if they occur, are also slower to begin. Perhaps lesser fluctuation under this maritime climatic regime tends to prolong the effects of individual seasons.

Colonization at Phinney Bay may be responding to competitive interactions on panels to a greater degree than at Elliott Bay. The decline in species numbers during the beginning of the second year is associated with an increase in percent cover at that time, lending credence to the possibility that interactions between colonists may result in species reductions.

Although equilibrium species numbers during much of the first year of colonization are remarkably similar (20.5 ± 0.7 at Site 2; 21.6 ± 1.1 at Site 1), these numbers are considerably greater than those determined in the

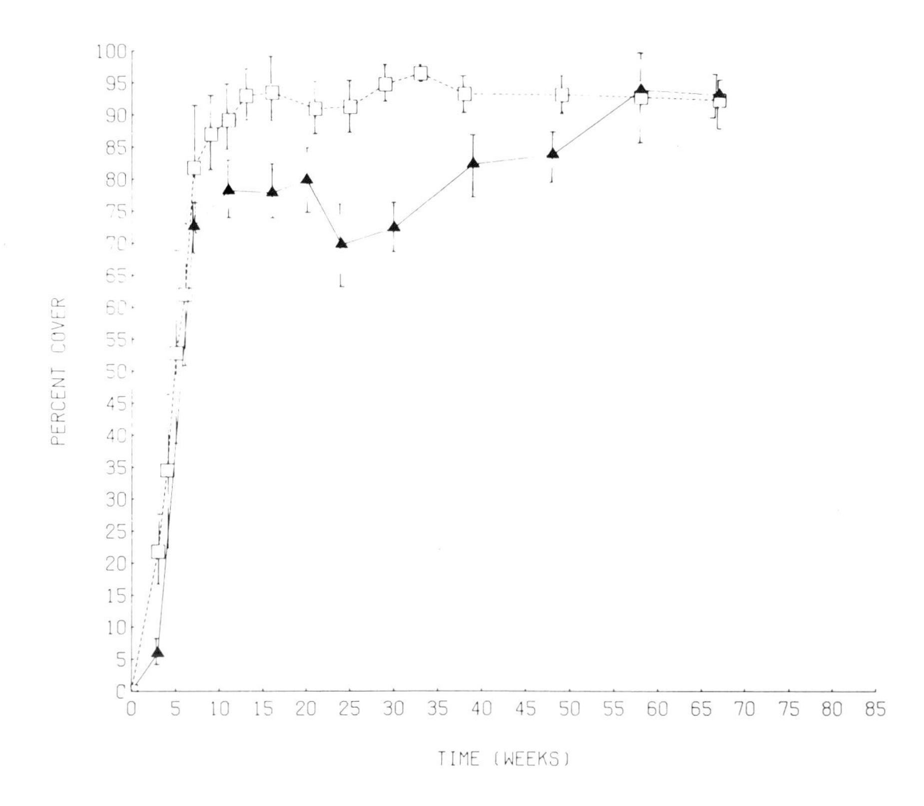

Figure 7. Mean percent cover of panels at Sites 1 and 2, plotted against time. Symbols as above.

interlatitudinal comparison [25]. This is partly attributable to differences in the depth of water and the distance from shore in those studies: the farther offshore and the greater the depth at which panels are positioned, the fewer species that accumulate [12]. As would be expected, these parameters were greater for the interlatitudinal comparison than for the present study.

Investigating the functional groups [37] of colonists that together comprise the total percent cover estimates given above suggests the presence of additional patterns. By classifying colonists as either solitary or colonial, in accordance with Jackson's [2] designations, several patterns emerge. Colonial organisms include, for example, a variety of bryozoans, hydroids and some species of tunicates, all of which can reproduce asexually as well as sexually. Solitary organisms include only sexually reproducing species, e.g., mollusks, barnacles and some other tunicate species. As Jackson [2] points out, there are several other basic differences between these types of space occupiers, including the ability of some colonial organisms (bryozoans excluded) to resist secondary fouling because of their soft-bodied surfaces, and the abilities of solitary organisms to outdistance colonial organisms by growing vertically above the substrate. Representatives of both types of space occupiers are present in Puget Sound throughout the colonization period. At both these localities, colonial organisms establish themselves immediately and dominate panels for considerable lengths of time during the first year of colonization. Figure 8 indicates changes in dominance between colonial and solitary species for the Elliott Bay fouling community. The same pattern is observed at Phinney Bay, although the species involved are different. By contrast, at tropical localities colonial organisms eventually predominate, but it takes considerably longer (14–26 months) for them to emerge as dominants over solitary species [2]. That colonial organisms predominate initially in Puget Sound suggests that they are both present in substantial numbers and that their asexual growth enables them to spread rapidly before solitary organisms present can enlarge. That this is less so in tropical areas is supported by Jackson's [2] Figure 1, which indicates that tropical panels are only 20% covered by attached growth after half a year, and approximately 50% covered after more than one-year's exposure. It takes from one to two years for tropical panels at his Jamaican locality to reduce the amount of free space available and to begin to achieve extensive coverage by colonial species. On Puget Sound panels at that time, colonial species become relatively less important space occupiers and, instead, solitary species begin to predominate. This happens earlier at one site than at the other, occurring after one year at Elliott Bay and after half a year at Phinney Bay.

Possible reasons for the early demise of the intially successful colonial species found at both Puget Sound localities may encompass several aspects of their life history. First, among some colonial species, e.g., bryozoans, a considerable fraction of their resources is devoted to expansionist policies; lateral growth may leave a central region in disrepair and a perhipheral region alive. If colonies can achieve greater diameters in Puget Sound, larger areas can become recolonized. Greater food availability or fewer similarly enlarging colonists per panel may enable such differences. Second, encrusting species expanding laterally may be physically dislodged by their attempts at overgrowth; this is true if adjacent species possess hard, calcareous skeletons and increase their dimensions before coverage can take place, encouraging a wedge effect that may disturb portions of the colonial organisms. Third, since the colony is interconnected, disturbance to its peripheral areas may result in damage to areas beyond those affected directly. Studying the undersurface of

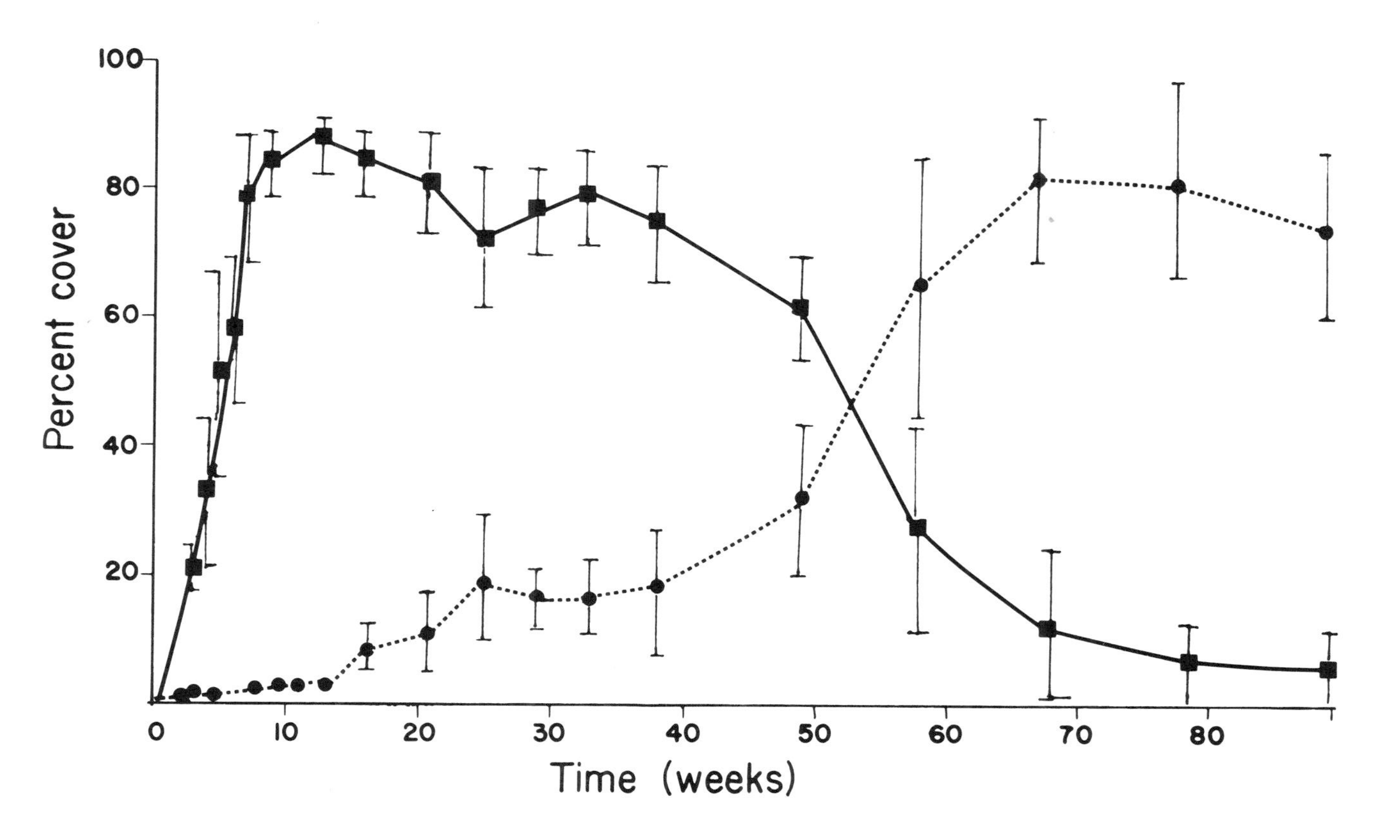

Figure 8. Mean percent cover of panels by colonial organisms (solid line) and by solitary organisms (dashed line) at Site 1, plotted against time. The 95% confidence intervals are given as bars.

panels may exacerbate this effect, although the ability to repair partially disturbed portions of the colony does exist [2]. Although a single disturbance to a colonial organism potentially could disrupt the entire colony, this need not occur. Fourth, the actual or realized longevities of colonial and solitary species may vary; with few data available there is ample room for speculation. Some data indicate that the same bryozoan species may exhibit differences in life spans of as much as eight years; Sutherland [22] contrasts the two-year life span of *Schizoporella* in North Carolina with its ten-year life span in Massachusetts. Seasonal regression of colony size may be as effective; some species, e.g., the colonial tunicate *Distaplia occidentalis,* decrease their colony size during winter. Solitary species can increase their coverage during this period if their larvae are available to settle.

DIFFERENCES IN SPECIES COMPOSITION BETWEEN SITES

As has been pointed out by researchers investigating the histories of panel communities [6,8], multiple outcomes or stable points are a reality when identical substrates are observed. That otherwise similar substrates develop along different paths is not only true for a particular locality, but occurs between localities as well. The experimental sites in Puget Sound acquired different dominant phyla during the first two years of colonization. Earlier studies at Phinney Bay indicate that tunicates were as important colonists a decade ago as now [38]. Other than differences in larval availability, which may or may not be related to submergence time or differences in the physical regimes of these areas, other explanations can account for these observations.

One possible explanation involves the effects of predation on panel systems. The relationship between predation and community structure has been approached primarily with respect to large predators, e.g., fish. By caging panels, this type of predation is effectively prevented. Sutherland [8] reports that soft-bodied tunicates prevail on panels covered by mesh, while encrusting hard-bodied bryozoans exist outside the mesh. Lacking skeletonized parts, tunicates and other soft-bodied prey might be ripe for removal. Similar results are observed on tropical fouling panels [39,40]. Interestingly, the types of colonists persisting outside enclosures are similar to those found at Site 1, and those inside enclosures similar to those found at Site 2 in Puget Sound. Predation intensities between localities may be implicated in maintaining some of the observed compositional differences.

Although predation undoubtedly is important in structuring natural hard substrate communities (see Paine [41] for review), there is some question of its influence on these isolated and small-sized substrates. Sutherland and Karlson [10], unable to duplicate grazing results on temperate panels, relegate predation to a less important position. Both Jackson [2] and Sutherland [22] dismiss predation as an unimportant influence on tropical fouling panels, although Day [39] reports otherwise. While cages prevent access by large predators, it is possible that small predators (e.g., nudibranchs and nemerteans) gain access to panels even though caged.

Other factors that could account for different outcomes relate to subtle initial settlement or growth differences between colonists at these localities. Early colonization at both include substantial hydroid coverage. In addition to providing an extensive protective canopy, hydroids may play an active role in subsequent settlement by other species. Several experiments attest to this possibility. Although cages encourage the growth of tunicate species by protecting them from possible predation, dense canopies of hydroids or erect bryozoans can be similarly effective [8]. Russ [40] reports equivalent effects generated by attaching artificial arborescent bryozoans to panels, suggesting that any mechanism that decreases the foraging efficiency of fish may effectively reduce predation. By manually clipping hydroid growth, Standing [42] observed significant differences in barnacle recruitment to substrates. He likewise found tunicates present in significantly greater abundances where hydroid growth was most dense, but related their abundance to preference of larvae for dark, protected places. All of these examples imply that biologically generated habitat complexity may be additional factors in determining species compositions on panels and may explain differences between sites.

CONCLUSIONS

These in-depth comparisons of the colonization process in two areas of Puget Sound indicate extensive similarities in many aspects of colonization although the actual identities of the species involved may differ. These similarities primarily concern the rates at which sessile colonists appear and their percent cover. Functional groups dominating over the time interval also show patterns. These observations strengthen the conclusion that colonization can be described quantitatively even when major species compositional differences are discernible. These parameters may serve as a base for estimating geographic patterns in colonization.

Competition between species and predation or other forms of disturbance on colonists, may affect the subsequent panel community structure, resulting in variation between panels at a locality or between localities. Despite a somewhat limited understanding of the mechanisms structuring these different panel communities, colonization can be portrayed in a quantitative fashion.

REFERENCES

1. Coe, W. R., and W. E. Allen. "Growth of Sedentary Marine Organisms on Experimental Blocks and Plates for 9 Successive Years at the Pier of the Scripps Institution of Oceanography," *Bull. Scripps Inst. Oceanog. Univ. Calif. Tech. Ser.* 3: 37–85 (1937).
2. Jackson, J. B. C. "Competition on Marine Hard Substrata: the Adaptive Significance of Solitary and Colonial Strategies," *Am. Nat.* 111: 743–767 (1977).
3. Long, E. R. "Marine Fouling Studies off Oahu, Hawaii," *Veliger* 17: 23–36 (1974).
4. Woods Hole Oceanographic Institution. *Marine Fouling and Its Prevention* (Washington, DC: U.S. Naval Institute, 1952).
5. Boyd, M. "Fouling Community Structure and Development in Bodega Harbor, California," Ph.D. Thesis, University of California, Davis, CA (1972).
6. Osman, R. W. "The Establishment and Development of a Marine Epifaunal Community," *Ecol. Monog.* 47: 37–63 (1977).
7. Osman, R. W. "The Influence of Seasonality and Stability on the Species Equilibrium," *Ecology* 59: 383–399 (1978).
8. Sutherland, J. P. "Multiple Stable Points in Natural Communities," *Am. Nat.* 108: 859–873 (1974).
9. Sutherland, J. P., and R. H. Karlson. "Succession and Seasonal Progression in the Fouling Community at Beaufort, North Carolina," *Proc. 3rd Int. Cong. Marine Corrosion and Fouling* (Evanston, IL: Northwestern University Press, 1973, pp. 906–929.
10. Sutherland, J. P., and R. H. Karlson. "Development and Stability of the Fouling Community at Beaufort, North Carolina," *Ecol. Monog.* 47: 425–446 (1977).
11. Blick, R. A. P., and R. Wisley. "Attachment Rates of Marine Invertebrate Larvae to Raft Plates at Sydney Harbour Site 1959–1963," *Aust. J. Sci.* 27: 84–85 (1964).
12. DePalma, J. R. "Fearless Fouling Forecasting," in *Proc. 3rd Int. Cong. Marine Corrosion and Fouling* (Evanston, IL: Northwestern University Press, 1973), pp. 865–879.
13. Haderlie, E. C. "Marine Fouling Organisms in Monterey Harbor," *Veliger* 10: 327–341 (1968).
14. Nair, N. B. "Some Observations on the Distribution of Bryozoans in the Fjords of Western Norway," *Sarsia* 3: 36–45 (1961).

15. Nair, N. B. "The Settlement and Growth of Major Fouling Organisms in Cochin Harbor," *Hydrobiologia* 30: 503-512 (1967).
16. Ralph, P. M., and D. E. Hurley. "The Settling and Growth of Wharf-Pile Fauna in Port Nicholson, Wellington, New Zealand," *Zool. Publ. Victoria Univ. Wellington* No. 19: 1-22 (1952).
17. Turner, R. D. "Wood-Boring Bivalves, Opportunistic Species in the Deep Sea," *Science* 180: 1377-1379 (1973).
18. Grassle, J. F. "Slow Recolonization of Deep-Sea Sediment," *Nature* 265: 618-619 (1977).
19. Schoener, A. "Experimental Zoogeography: Colonization of Marine Mini-Islands," *Am. Nat.* 108: 715-738 (1974).
20. Sousa, W. P. "Experimental Investigations of Disturbance and Ecological Succession in a Rocky Intertidal Algal Community," *Ecol. Monog.* 49: 227-254 (1979).
21. Strathmann, R. R., and E. S. Branscomb. "Adequacy of Clues to Favorable Sites Used by Settling Larvae of 2 Intertidal Barnacles," in *Reproductive Ecology of Marine Invertebrates,* S. Stancyk, Ed. (Columbia, SC: University of South Carolina Press, 1979), pp. 77-90.
22. Sutherland, J. P. "Dynamics of the Epibenthic Community on Roots of the Mangrove *Rhizophora mangle* at Bahia de Buche, Venezuela," *Mar. Biol.* 58: 75-84 (1980).
23. Zann, L. P., R. J. Cuffey and C. Kropach. "Fouling Organisms and Parasites Associates with the Skin of Sea Snakes," in *The Biology of Sea Snakes,* W. A. Dunson, Ed. (Baltimore, MD: University Park Press, 1975), pp. 251-266.
24. Wells, H. G., N. J. Wells and I. E. Gray. "The Calico Scallop Community in North Carolina," *Bull. Mar. Sci. Gulf Caribb.* 14: 561-593 (1964).
25. Schoener, A., E. R. Long and J. R. DePalma. "Geographic Variation in Artificial Island Colonization Curves," *Ecology* 59: 367-382 (1978).
26. Green, R. H. *Sampling Design and Statistical Methods for Environmental Biologists* (New York: John Wiley & Sons, Inc., 1979).
27. Chieppa, M., A. Barbara, A. Francescon, G. Relini and A. Tursi. "Le repliche nello studio del fouling," *Ann. Istut. Stat.* 38: 1-25 (1979).
28. Mook, D. "Studies of Fouling Invertebrates in the Indian River," *Bull. Mar. Sci.* 26: 610-615 (1976).
29. Cairns, J. Jr., and W. H. Yongue, Jr. "The Distribution of Fresh-Water Protozoa on a Relatively Homogeneous Substrate," *Hydrobiologia* 31:65-72 (1968).
30. Cairns, J. Jr., J. L. Plafkin, W. H. Yongue, Jr. and R. L. Kaesler. "Colonization of Artificial Substrates by Protozoa: Replicated Samples," *Arch Protistenk Bd* 118: 259-267 (1976).
31. Schoener, A., and C. H. Greene. "Variability Among Identical Fouling Panels in Puget Sound, Washington, USA," *Proc. 5th Int. Cong. Marine Corrosion and Fouling,* Barcelona, Spain, (1980), pp. 213-224.
32. Siegel, S. *Nonparametric Statistics for the Behavioral Sciences* (New York: McGraw-Hill Book Co., 1956).
33. English, T. S. "A Theoretical Model for Estimating the Abundance of Planktonic Fish Eggs," *Conseil Perm. Int. l'Expl. mer* 155: 174-182 (1964).

34. Sutherland, J. P. "Function Roles of *Schizoporella* and *Styela* in the Fouling Community at Beaufort, North Carolina," *Ecology* 59: 257–264 (1978).
35. MacArthur, R. H., and E. O. Wilson. *The Theory of Island Biogeography* (Princeton, NJ: Princeton University Press, 1967).
36. Simberloff, D. S., and E. O. Wilson. "Experimental Zoogeography of Islands: the Colonization of Empty Islands," *Ecology* 50: 296–314 (1969).
37. Woodin, S. A., and J. B. C. Jackson. "Interphyletic Competition Among Marine Benthos," *Am. Zool.* 19: 1029–1043 (1979).
38. Lambert, G. "The General Ecology and Growth of a Solitary Ascidian, *Corella willmeriana,*" *Biol. Bull.* 135: 296–307 (1968).
39. Day, R. W. "Two Contrasting Effects of Predation on Species Richness in Coral Reef Habitats," *Mar. Biol.* 44: 1–5 (1977).
40. Russ, G. R. "Effects of Predation by Fishes, Competition and Structural Complexity of the Substratum on the Establishment of a Marine Epifaunal Community," *J. Exp. Mar. Biol. Ecol.* 55–69 (1980).
41. Paine, R. T. "Controlled Manipulations in the Marine Intertidal Zone and Their Contributions to Ecological Theory," *Acad. Nat. Sci. Spec. Publ.* 12: 245–270 (1977).
42. Standing, J. D. "Fouling Community Structure: Effects of the Hydroid *Obelia dichotoma* on Larval Recruitment," in *Coelenterate Ecology and Behavior,* G. O. Mackie, Ed. (New York: Plenum Publishing Corp., 1976), pp. 155–164.

CHAPTER 2

INTERACTIVE AND NONINTERACTIVE PROTOZOAN COLONIZATION PROCESSES

John Cairns, Jr. and Michael S. Henebry

Department of Biology and
University Center for Environmental Studies
Virginia Polytechnic Institute and State University
Blacksburg, Virginia 24061

INTRODUCTION

For many years, microbial ecologists thought that protozoan communities might be random aggregations of species. However, in 1937 Picken [1] showed that they have a structure, and in the last 30 years a substantial amount of information has accumulated that enables at least a preliminary observation that the colonization process is, as Simberloff [2] suggested, composed of two portions—an initial noninteractive stage followed by a longer interactive stage. Cairns and Yongue [3] and Cairns et al. [4] have discussed the possibility that the log-normal distribution first hypothesized by Preston [5] and subsequently verified by Patrick et al. [6] could be the result of random introduction of species into environments of different degrees of suitability. The majority of invading species would find the environment hostile or unsuitable of course, but the ones remaining might find the degrees of suitability covered in Figure 1. Cairns [7] examined the possibility of synergistic interactions among species of protozoans, and this hypothesis was elaborated further by Cairns and Yongue [3]. They hypothesized a fixed structure based on functional units and their interactions with other functional units (Figure 2). These hypotheses are not verified easily because of the difficulty of studying a natural microbial community without interfering with its processes, and

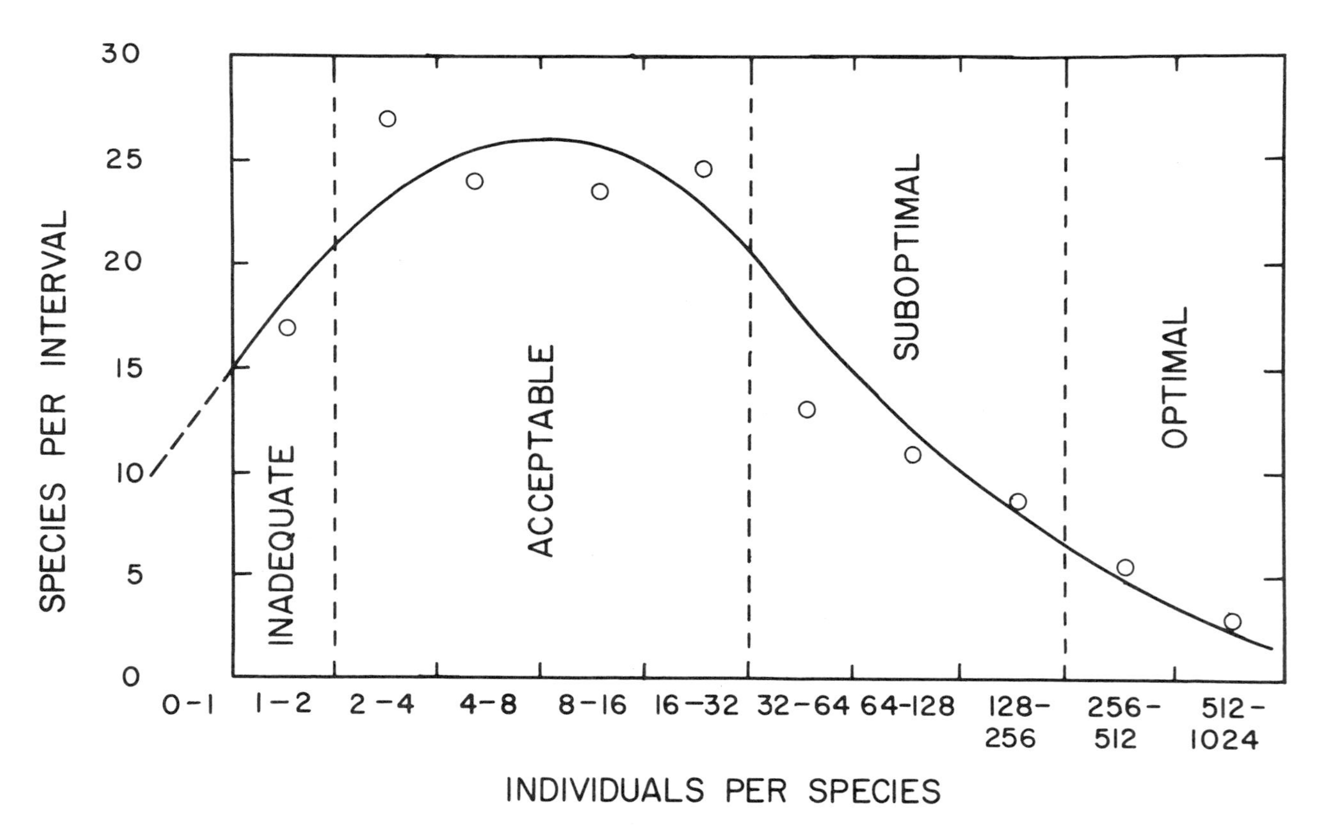

Figure 1. A model for the truncate normal curve distribution. The four zones relate to conditions for survival and colonization.

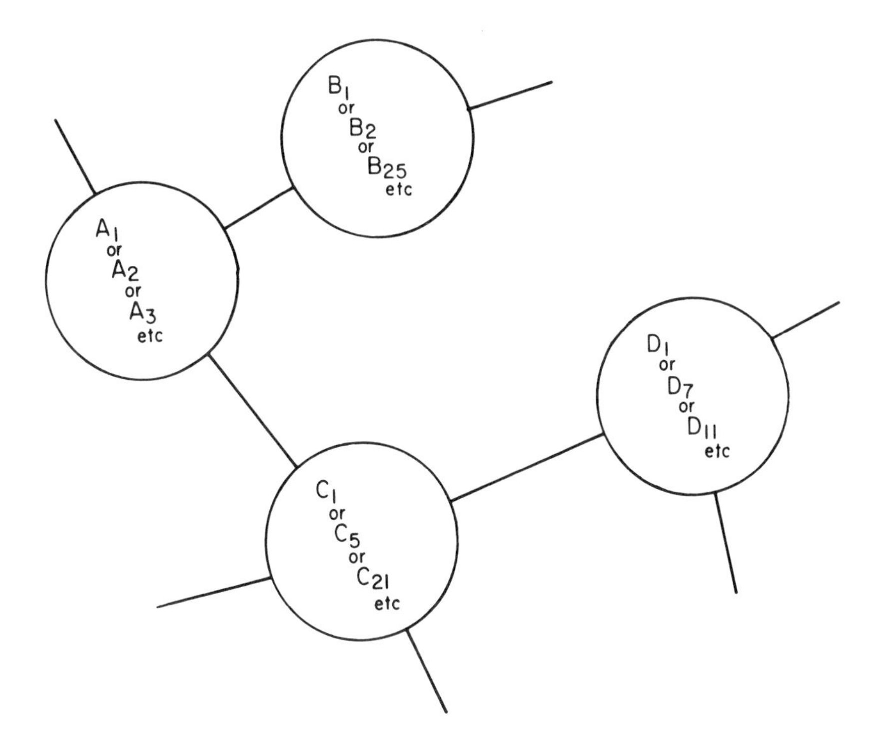

Figure 2. Functional role alternation by protozoan species. Role A might be filled temporarily by a species designated A_1 or, as far as species temporarily filling role B is concerned, A_7. Thus, community structure would be relatively stable despite species replacing each other because the "role relationships" remain constant. Division of role A among species might be determined by differential tolerances to pH, temperature, etc. The number of roles would, of course, greatly exceed those depicted in this simplified diagram.

also because of the high degree of variability characteristic of natural systems. The purpose of this chapter is to examine the colonization process using existing evidence to determine which portions are likely to be noninteractive and which portions are likely to be interactive.

OPPORTUNITIES FOR SPECIES–SPECIES INTERACTIONS

Free-living protozoans appear to have a cosmopolitan distribution and appear wherever and whenever suitable ecological conditions exist. Thus, the species found in the upper Amazon drainage basin [8] are similar to those found in the Savannah River basin in the United States, which has a somewhat similar ecological environment [9]. This distribution is substantiated further by the fact that keys produced in Germany [10] are equally useful in the United States, South America and Asia. This utility of taxonomic keys and species descriptions also indicates that relatively little morphological change has occurred in these species since the invention of the microscope because the species descriptions produced many years ago are quite valuable today. Finally, Patrick [11] has shown that the numbers of species (of a variety of taxonomic groups, including protozoans) established in rivers located in a number of different geographic areas and having dissimilar water qualities are remarkably similar.

Most freshwater habitats are temporary, and a species of microorganism is faced with the problem of dispersing to the next body of water or of waiting until the next body of water comes to it (e.g., rain or floods). An adaptation that suffices to spread something as small as a protozoan from one lake to another probably suffices to spread it from one end of the earth to the other. Consequently, any given lake or river is probably subject to invasion by a very wide selection of microorganisms. Moreover, any organism as small and short-lived as a protozoan is extraordinarily sensitive to environmental change; indeed, the changes that take place from week to week in one lake may be as important to it as the differences between one lake and another lake.

Putting these two factors together, one might expect that as the environment changes a given protozoan species is quite likely to be replaced by one of the ever-present series of invaders of its habitat best adapted to the changed conditions. Therefore, we are confronted with that succession, so strikingly characteristic of protozoan communities, of protozoan species rising to abundance only to vanish. As a consequence, most free-living protozoans must be fugitive species.

Some protozoans modify their environment in a manner beneficial to others [12-16]. One may expect natural selection to have shaped these interactions into synergistic associations or patterns of species. Once such a synergistic pattern develops for a given type of habitat, it is advantageous for a would-be invader of that habitat to be able to take advantage of the pattern so as to be able to fit into it. To be associated with another pattern is less useful to the organism because successful invasion then would depend on the simultaneous invasion of several species. Thus, because protozoans are fugitive species, we would expect that once such a synergistic pattern developed for a given type of habitat that the pattern would tend to become nearly universal for that type of habitat.

Further, we would expect that it would have acquired the property that each role or niche in it could be occupied by any one of a number of species. The species found to occupy that niche in a particular community at a given time would depend on some combination of environmental conditions and historical accident.

That protozoans are one of the oldest groups of organisms on the earth, coupled with their cosmopolitan distribution and morphological stability, have provided ample opportunity for particular species to be associated with each other. However, Cairns [17] found that the frequency of pair associations at a particular locus in time and space was not as common as this hypothesis suggests.

An analysis was made of the results obtained by the Academy of Natural Sciences of Philadelphia river-survey team. The data were obtained from a series of 202 areas in various rivers and streams, mostly in the United States but also from some other parts of the world. All these areas were classified as healthy or semihealthy according to the system described by Patrick [18,19]. Although most of the sampling took place in the spring, summer and fall, a number of winter samplings are included in this total. Each sampling area included approximately 100–300 meters of stream or river from bank to bank, and commonly included a riffle, slack water and pool area. Collections were obtained in the morning from all common habitats, such as algal growths, mud surfaces and submerged roots, in both shaded and sunlit areas, and were placed in half-pint, screw-top jars, leaving a half-inch air space above the water surface. These samples were carried immediately to the field laboratory where identifications were made from living specimens. Generally about 20 half-pint samples were collected, and approximately a dozen different examinations made from each jar.

The samples first were examined under a low-power lens to determine the general structure of the population. More critical and detailed examinations followed. Most samples were taken from the meniscus and the bottom of the jar, both toward and away from the principal source of light, although an effort was made to ensure that attached species (to twigs, leaves, algal filaments, etc.) were not overlooked. Before the sampling began, several concentration techniques were tried and discarded because many species were destroyed or injured and many streams had a comparatively heavy silt load, which interfered with both concentration and examination. Only those species with a density of at least six individuals or more per drop of water from at least one sample were recorded. Furthermore, no samples of the plankton are included in this report, but only those species in association with the substrate. Nearly 1200 species were identified in the course of these studies, of which approximately 25% occurred in four or more of the areas sampled. The 1200 species represent only a fraction of those described. For example, nearly 6000 species of ciliates alone, most of which are free-living, have been described to date according to Corliss [20]. Three-quarters of the 1200 species recorded in this series of samples occurred less than 1.6% of the time.

Of the 25% of the species that occurred four or more times in the 202 areas sampled, only 20, or roughly 6%, occurred in at least 25% of the areas studied. The number of areas in which each of these 20 species occurred was as follows: 125, 87, 80, 69, 67, 66, 64, 64, 62, 60, 60, 60, 55, 53, 52, 49, 49, 48, 47, 47. Naturally, these ubiquitous species occurred together quite often. In an effort to determine whether pairs of these species occurred together more frequently than would be expected from chance alone, an association matrix was made for the 20 sampling areas in which these species were most common. A Chi-square test of significance was run on the 190 possible associations of species pairs and, of these, 44 occurred together more frequently than expected from chance alone at the 5% level of confidence. Also, when the Chi-squares for all of the paired associations were summed, the result showed a highly significant divergence from chance expectation. Comparison of the expected and observed frequency distributions of Chi-square values showed that the divergence was due to the aforementioned large number of significant associates at the 5% level of confidence. An examination of the data indicated that associations of three or more species also existed. That these species were able to occur together suggested similar environmental requirements and that the environment might be the major factor producing these associations. To check this possibility, the environmental data were processed with a computer and information was obtained for each of the 20 most common species throughout the entire range of occurrences.

When the results for the 20 most common species were compared, three factors were noted: (1) pairs of larger groups of associated species always had virtually identical ranges of environmental conditions; (2) these species always tolerated rather broad ranges of environmental conditions; and (3) having identical ranges of tolerance to the chemical and physical environment does not ensure that species will be associated more often than would happen by chance alone, since only 44 pairs out of 190 were associated significantly, although all species had broadly overlapping environmental ranges.

These results do not eliminate species-to-species interactions, but the results indicate that beneficial interactions that require the simultaneous presence of particular morphological species will occur only a small percentage of the time that either of the species is occupying a habitat. Therefore, it is quite likely that both competitive and synergistic or beneficial interactions are much more likely to be based on the ecological role than on the morphological characteristics.

The research just described was carried out with natural substrates and provides base data from which concepts have been developed for the studies using artificial substrates.

Description of Artificial Substrates

Although a wide variety of artificial substrates have been used for studying microbial colonization [21], the ones described in these studies as polyurethane foam (PF) units are the ones we found most satisfactory. We emphasize this series of studies because it represents the largest body of data involving freshwater protozoans and artificial substrates availble for analysis.

Polyurethane forms an open-cell foam that resembles an intricate latticework of pillars and interstitial spaces under the scanning electron microscope (Figure 3). The three-dimensional character of the foam permits ready colonization by a wide variety of microorganisms; free-swimming forms easily invade the interstices of the foam, while sessile forms attach to the solid pillars [22]. It is essentially an inert substance, which can be autoclaved without altering its effectiveness as a sampling device [23], although this was not done except for the study just cited.

Field studies have shown that most protozoan species found by directly sampling the natural environment eventually colonize PF units [24]. In fact, polyurethane foam seems to be an attractive generalized substrate for protozoa from virtually every microhabitat. Natural substrates are more selective and usually collect fewer species [25]. Thus, this inert, nonselective, physically complex substrate material that allows a wide variety of organisms to colonize on it can be used effectively to integrate sampling over the various microhabitats of a study area. Small quantities of polyurethane foam can be purchased from almost any department store, or it can be obtained in bulk from Bernal Foam Products, Inc., Buffalo, New York.

Replicability of Samples from PF Units

In a study of two Michigan ponds, Yongue et al. [26] found a high degree of species overlap among replicate PF units. To further investi-

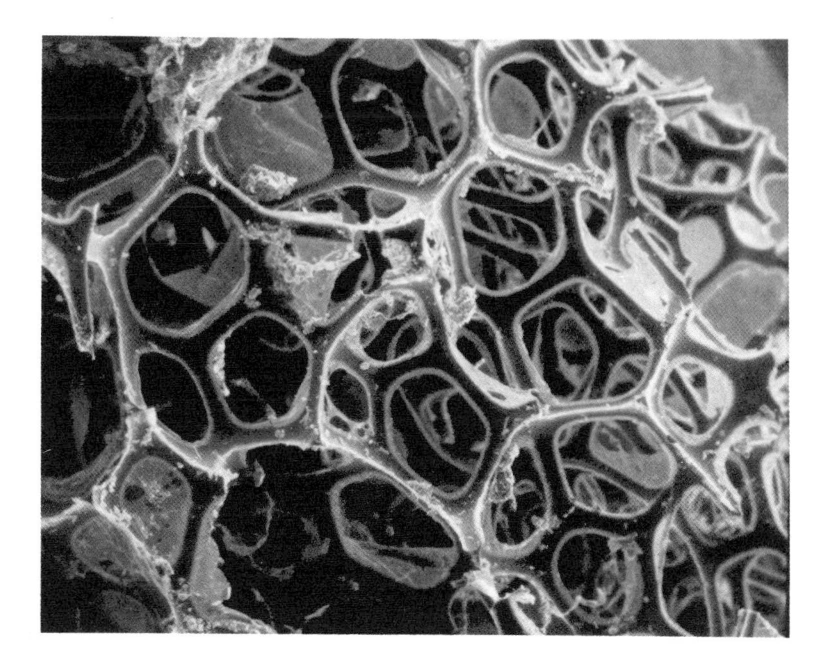

Figure 3. Scanning electron micrograph of polurethane foam material exposed for 10 days in New River, Virginia [22].

gate the replication properties of these substrates, Cairns et al. [27] placed two series of 10 PF units each at two similar sites in Douglas Lake, Michigan (some others were placed at ecologically quite different sites as well). After three weeks of immersion, the substrates were harvested and the colonizing protozoa identified. Cluster analysis (see dendrogram Figure 4, clusters B and C) indicated high correlation among replicate substrates of a 10-unit set. Correlation between sets was also high, indicating that PF units drew essentially comparable samples from different locations of similar water quality.

These results differed substantially from those found by Patrick et al. [29] in a study in which natural substrates were sampled. Despite their geographical proximity and similar water qualities, sampling stations in that study showed very little overlap in protozoan species composition. Comparing these seemingly contradictory results graphically illustrates the advantages of using artificial substrates to enhance replicability and ensure acquisition of comparable samples from different survey locations. It is quite likely that differences in microhabitat quality not easily seen were responsible for the low overlap in their study.

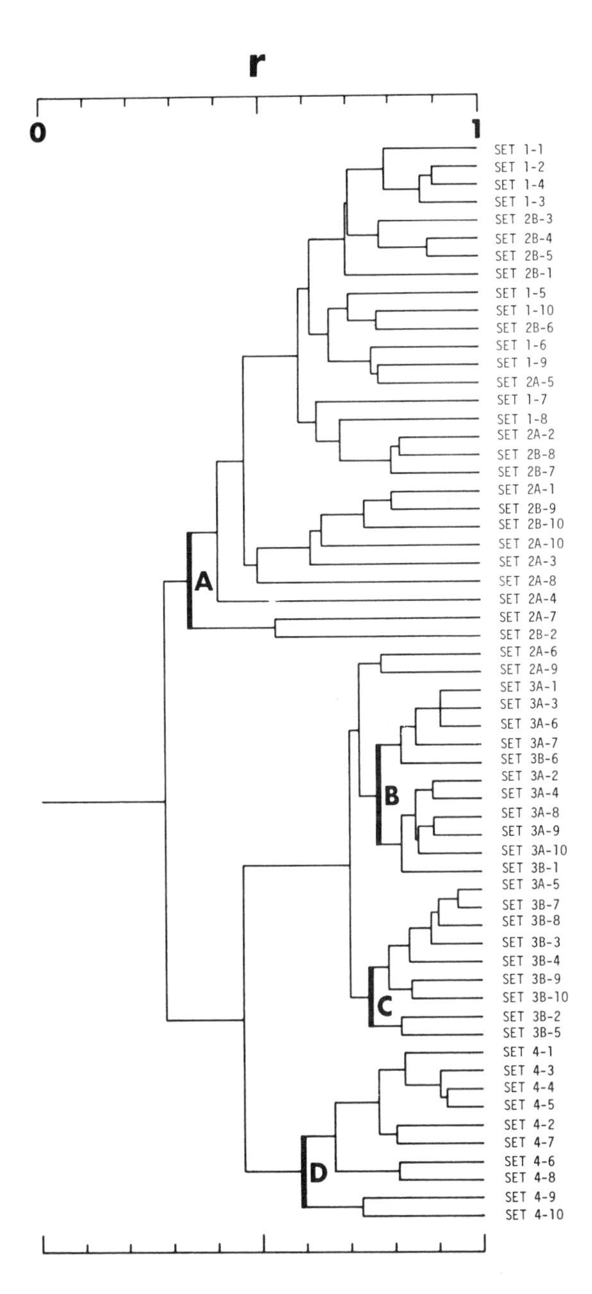

Figure 4. Q-mode dendrogram prepared by the unweighted pair-group method with arithmetic averages [28] from a matrix of correlation coefficients [27]. Sets 3A and 3B are two sets of 10 replicate sponges colonized for 3 weeks in Douglas Lake, Michigan.

ARTIFICIAL SUBSTRATES AS HABITAT ISLANDS

Generally, artificial substrates can be thought of as initially barren habitat islands with the natural community acting as a source pool of potential colonists. Three factors that critically influence the colonization of artificial substrate islands and, hence, their effectiveness as sampling devices are: (1) the size of the substrate; (2) the nature of the substrate material and its position in the ecosystem; and (3) the length of time the substrate is exposed to colonization.

Substrate Size

Island size is unquestionably a critical factor in determining the number of species that can exist on an island. Whether size functions directly to enchance species number or indirectly by increasing habitat diversity [30] is not totally clear, but its gross effect on species number should be considered when utilizing an artificial substrate as a sampling device. The substrate should be of sufficient size to allow colonization by a large number of species and ensure that few ecologically important species are initially excluded.

Substrate Position

Simberloff [2] has shown that "any patch of habitat isolated from similar habitat by different, relatively inhospitable terrain traversed only with difficulty to organisms of the habitat patch may be considered an island." A spatially stable substrate, such as a PF unit positioned in the open waters, could be considered an island, since most of the organisms adapated to survival in such a habitat (Aufwuchs species) must traverse "inhospitable terrain" to colonize it. Planktonic species, although less well adapted to a substrate existence, may dominate the pioneer episodes of colonization if substrates are suspended in open water. Only subsequently do organisms from the comparatively distant solid substrates invade. Thermal stratification [31], prevailing currents [32] and chemical gradients [33] also can alter dispersal patterns and, thus, the nature of the substrate assemblages. By contrast, a PF unit placed in the porous sediments closely mimics the contiguous natural substrate [34].

Well-adapted species readily invade from the immediate surroundings and quickly dominate the PF unit community. Depending on position, substrates behave as habitat islands that closely resemble the adjacent environment; or, if distant from similar microhabitat, they behave as habitat islands that during the course of colonization are affected by the total species pool of the larger system.

Length of Exposure to Colonization

Initial island colonization is essentially a noninteractive process primarily influenced by the dispersal capacities and extinction potentials of the colonizing organisms [35]. With the establishment of an equilibrium species number, interactive processes such as competition and predation take precedence in determining the island's species composition [36]. The island assemblage soon manifests the characteristics of an autonomous community capable of maintaining its integrity in the face of environmental change [34].

During the noninteractive phase of colonization, artificial substrates act as sampling devices that passively collect organisms from the natural community. With the acquisition of equilibrium, the substrate ceases to function primarily as a sampling device, and the associated species assemblage begins to evolve its own characteristic composition. If an artificial substrate is to be used as a species sampling device, it should be retrieved from the environment before the number of species attains equilibrium. The appropriate immersion time will vary under different environmental conditions, making comparable sampling days difficult to choose.

Because the effectiveness of an artificial substrate sampler can be affected differentially by the factors of substrate size, position in space and immersion time in a particular environment, it is preferable to (1) hold size constant when comparing specific systems; (2) position substrates so that they act as habitat islands; and (3) monitor colonization through time as an alternative to one-time sampling. A time-dependent process is then analyzed as the indicator of natural community structure.

Placement of Substrates and Collection of Samples

To assure that the substrates are of sufficient size to collect the majority of species from a location, PF units $5 \times 7.5 \times 6.5$ cm in size are used routinely [37]. Although variations in pore sizes and colors among lots of foam material do not appear to affect the numbers and kinds of colonizing protozoan species [25], it is probably best to cut all PF units for a study from a single lot. Specific units to be placed in a given location then are selected randomly from the batch. Units are attached to nylon rope with monofilament line or string and spaced so that they will not become entangled (approximately 30 cm apart).

The line is fastened to floats and suspended near or under the surface (Figure 5). This position will allow the substrates to behave as habitat islands and permit colonization by most littoral, benthic and pelagic protozoa. Spoon [38] has pointed out that the porous nature of these substrates probably excludes larger macroinvertebrate predators from the internal areas of the PF units. This merely strengthens the analogy between foam units and natural islands, since higher-order predators are selectively excluded from true oceanic islands as well [2]. As long as the units are near the surface of the open water and not in the immediate vicinity of natural solid substrates, exact depth of the water column is not crucial. All PF units in a sampling location should be randomly assigned predetermined harvesting dates to eliminate any systematic variation due to position on the line.

Prior to analyzing PF units from a location, the investigator should familiarize himself with the local protozoan fauna by inspecting a few composite samples from surrounding natural substrates. (Cairns [39] gives a detailed discussion of sampling.) Many of the more difficult identifications then can be clarified before quantitative analysis begins.

At a given harvesting, units are cut from the suspension line and

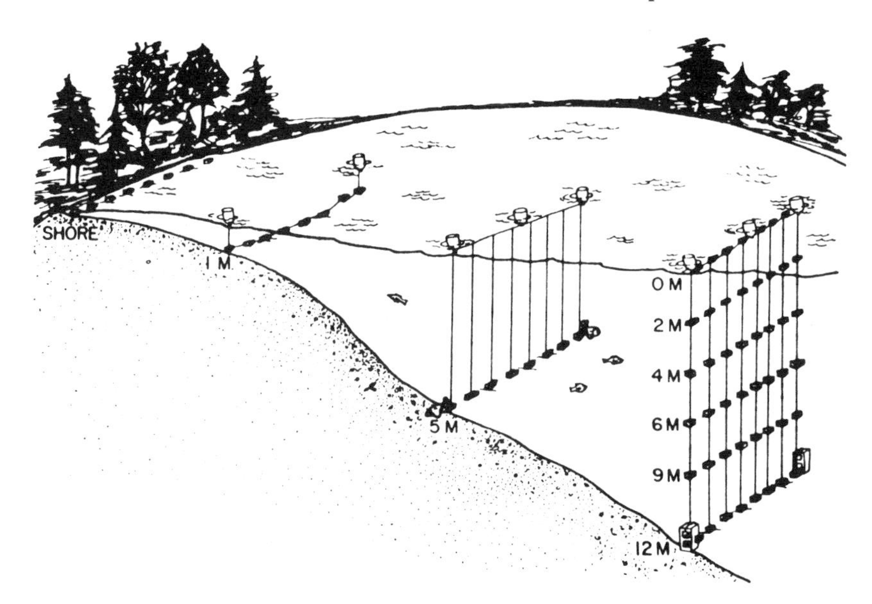

Figure 5. Anchoring substrates at various depths.

placed in individual containers with approximately 100 ml of surrounding surface water. Plastic Whirl-pak bags are compact, presterilized and easy to carry, but they are somewhat susceptible to puncture. Screw-capped jars also can be used, although they are comparatively difficult to transport. Four replicate substrates are generally harvested on a given day; samples should be transported to the laboratory as soon as possible in insulated cartons. On arrival, the containers are opened to allow free gaseous exchange. Units that cannot be analyzed immediately (the most desirable approach) should be stored at the ambient temperature of the sampling location.

At this point, the investigator must be conscious of the time limitations placed on the analysis of field samples. Even protozoa maintained in the environmental conditions in which they were collected are likely to die or reproduce very quickly and distort the assemblage structure. Fixation is not a practical solution to this problem because few protozoa retain morphologic integrity when fixed with conventional preservatives and, once fixed, are virtually impossible to separate from detrital material. Therefore, all initial identifications must be made on vital samples as soon as possible after collection.

Sample Analysis: Identification of Living Specimens

Approximately one hour before a particular substrate is to be analyzed, the excess water is decanted and as much material as possible squeezed from the unit into a sample jar. The squeezings then are placed in a fixed position in the laboratory where the organisms can respond to a single light source; some species tend to settle and accumulate on the bottom while others aggregate near the meniscus. It is not advisable to use filters to concentrate specimens because such procedures also concentrate debris and toxicants that can obscure and injure many protozoa [39].

Careful analysis of three to four wet-mount slides prepared from bottom and meniscus of the squeezings generally is adequate to define the species composition of a PF unit sample [40]. If an asymptotic number of species is not attained, additional slides must be examined (see Peet [41] for a discussion of sampling curves).

Determining the composition of field samples is probably the most laborious task confronting an investigator, and the accurate identification of specimens is an important aspect of community analysis. Some taxonomists feel that "precise cytological staining is a prerequisite to proper identification of species" [42], but if absolute systematic precision is demanded, ecological field studies can become interminable laboratory exercises. The use of the operational taxonomic unit, for example, has become a necessary standard practice in ecological work [43]. A pragmatic approach to the identification of living specimens is necessary if one is to study community dynamics.

As discussed earlier, delineation of protozoan community composition can be accomplished only if identifications are made on living material. Fixation and staining can be utilized to confirm preliminary identifications, but such techniques are not effective when organisms are in low density and mixed with natural debris. A method recommended by Spoon [44] for closely observing microorganisms under the light microscope is sometimes a useful alternative to staining.

By familiarizing himself with a few protozoology keys (bibliography), an investigator soon can develop confidence in the essential accuracy of his identifications on vital specimens. Cairns [39] has discussed techniques for maintaining consistency when working with living material, and a system for consolidating taxonomic references into a compact microfiche library [45] has been suggested as an aid. With experience, most commonly occurring protozoa can be identified accurately to at least the genus, if not the species, level. Gaining practical experience with field samples permits generation of ecologically pertinent and statistically useful information that otherwise would be impossible to obtain.

MICROECOSYSTEM STUDIES OF THE COLONIZATION PROCESS USING PROTOZOAN COMMUNITIES

Closed microecosystems (as defined by Cooke [46]) were used (1) to obtain baseline data on the colonization of islands by protozoa in a laboratory system for comparison with data collected in natural systems; (2) to investigate effects of island area and distance from a source pool of protozoan species (the latter almost impossible to observe in a natural situation) on the process of colonization; and (3) to test the effect of epicenter community development on the colonization process of surrounding islands [47].

Sets of three 72- × 64- × 50-mm polyurethane foam substrates were removed at weekly intervals from the benthic region of Douglas Lake and brought to the laboratory. Substrates used in the first two experiments had been in the lake for one year. Those used for the last three experiments had been in place for two, three and four weeks, respectively. All substrates were assumed to be fully colonized. Yongue and Cairns [24] and Cairns et al. [48] demonstrated that protozoan communities in Douglas Lake attained relatively stable equilibrium populations after 14 days under similar seasonal conditions. At each collection, one PF unit substrate was harvested and examined for numbers of species. These served as representatives of the epicenter communities. Each of the two PF units was placed intact in the center of a colonization tank filled with pasteurized water from Douglas Lake (Figure 6).

RESULTS – DISTANCE EFFECT

The number of species present on epicenter substrates was relatively constant for the five experiments, varying from 47 (experiment 4) to 54 (experiment 3). Distribution of the species present between the three classes of protozoans was also relatively constant – flagellates and ciliates typically composed 80–90% of the total and were represented by approximately equal numbers of species (Table I). Harvests of the experimental epicenters following experiments 2 and 5 revealed no substantial changes in the number of species present.

Figure 7 illustrates the accumulation of species with time on PF unit islands located different distances from the epicenter. Islands located 9.5 cm from the epicenter accumulated species faster and reached higher species numbers by the end of 7 days than islands of equal volume, but greater distance from the epicenter.

To determine whether the process of species accumulation on the islands was described adequately by the MacArthur-Wilson noninteractive island colonization model [$S(t) = \hat{S}eq(1 - e^{-GT})$ where $\hat{S}eq$ = equilibrium species number and with constant G related to transition time], nonlinear regression procedures with Marquardt methods of estimation [49] were used to fit the model to the experimental data. Lack of fit (L.O.F.) tests were used to test for any significant L.O.F. [50]. The island colonization model adequately described the accumulation of species at all distances (Table II) and showed no significant L.O.F. ($\alpha(F) > 0.01$).

The estimated equilibrium number of species ($\hat{S}eq$) for the nearest island (9.5 cm) was 23.43. Estimated equilibrium number decreased progressively for islands located 19 and 38 cm from the epicenters. The confidence ranges (calculated from standard errors) of estimates of $\hat{S}eq$ for the nearest and farthest islands did not overlap at the 90% level, although they did overlap with the range for the islands located at the middle distance. For the nearest islands it was estimated that 90% of the estimated equilibrium number ($t_{90}\%$, suggested by MacArthur and Wilson [51]) would be reached after 13.54 days. The $t_{90}\%$ for both sets of more distant islands was 7.19.

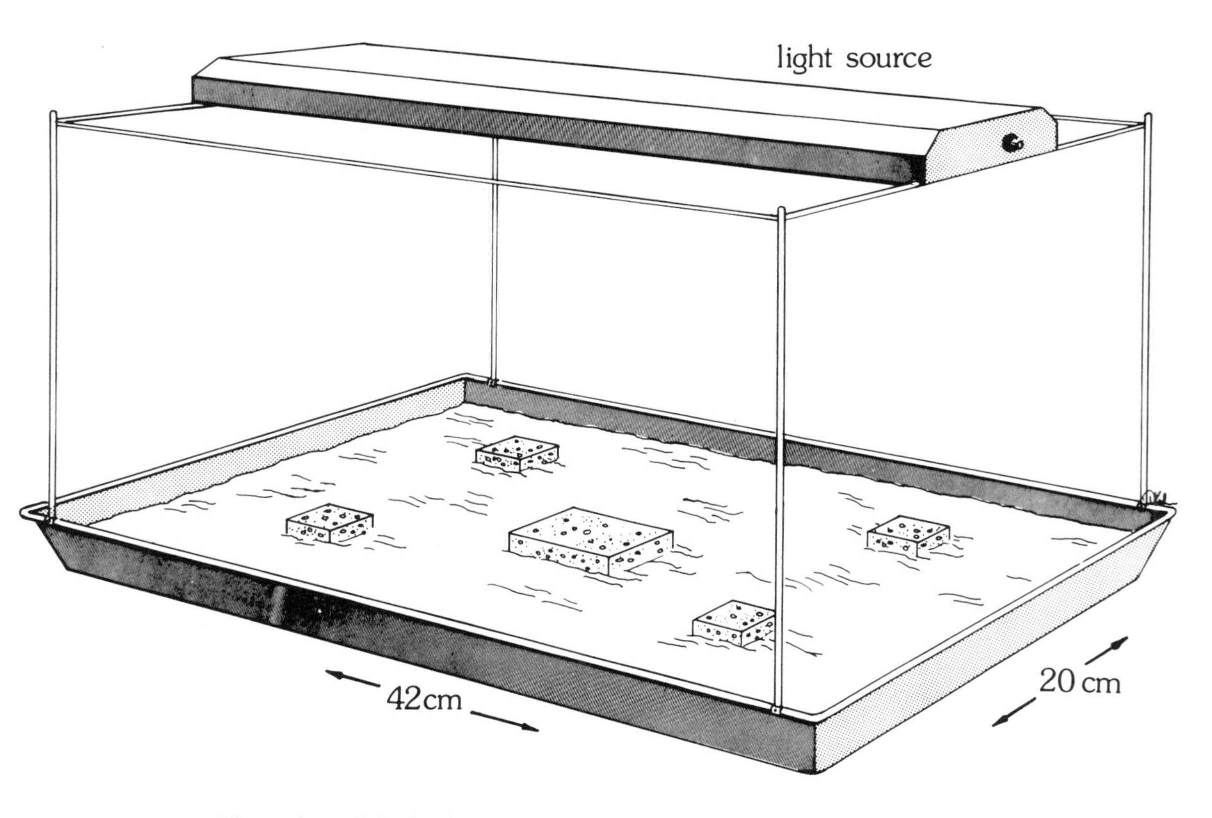

Figure 6. Colonization tank used in all microecosystem experiments.

Table I. Epicenter Species Richness

No. of Experiment	1	2	3	4	5
Colonization Time of Epicenter	1 year	1 year	2 weeks	3 weeks	4 weeks
Total No. Flagellates	25	22	22	18	23
Total No. Sarcodines	2	4	9	7	9
Total No. Ciliates	23	25	23	22	21
Total No. Species	50	51	54	47	53

Discussion of Distance Effect

The MacArthur-Wilson equilibrium model predicts that islands nearer to a source pool of colonists will equilibrate at a higher number of species than far islands of equal size. Additionally, it has been suggested that colonization on far islands proceeds more slowly [51]. Although field studies of the colonization of artificial islands by microorganisms [32,52,53] and small invertebrates [54] have confirmed other predictions of the species equilibrium model, their success in demonstrating the predicted distance effects has been minimal.

As noted by Schoener [54], the lack of success in field studies of the distance effects is the difficulty in defining the location and determining the composition of the source of potential colonists. Schoener's plastic sponge islands probably collected organisms from sources other than the designated epicenter (an algal mat) and, thus, blurred the effect of distance. Perhaps the best confirmation of distance effect using small aquatic organisms was that of Maguire [55]. In his Colorado study of the colonization of water-filled beakers, Maguire found decreased species richness as distance from a pond was increased. However, the influx of propagules from extraneous sources was again uncontrolled, and this may have influenced the results at the greater distances. In our laboratory experiments, the influx of organisms from sources other than the designated epicenters was minimized.

The parameters that were estimated as a result of fitting the model $S = \hat{S}eq(1 - e^{-GT})$ to the data were in accord with the predictions of MacArthur and Wilson [51]. The estimated $\hat{S}eq$ for the PF unit laboratory islands decreased with increasing distance from the experimental epicenters. The ranking of times to reach 90% of equilibrium number of species ($t_{90}\%$) did not follow the theory. The estimates from the fitted model indicated that the islands farther from the epicenter (38, 19 cm) would reach $t_{90}\%$ in about half the time of the nearest islands. However, the colonization curves (Figure 7) show that through day 7 the actual

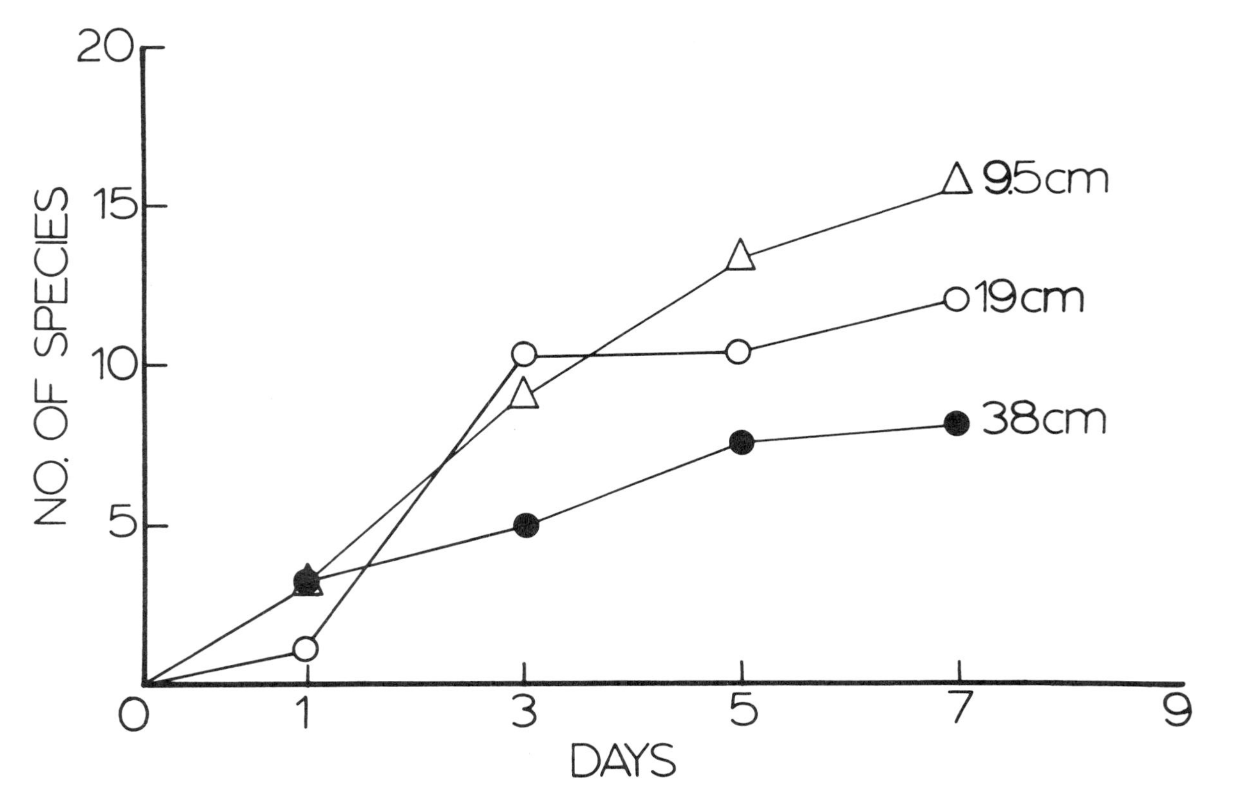

Figure 7. Colonization curves for equal-sized islands located at different distances from epicenters in laboratory micro-ecosystems.

Table II. Nonlinear Regression Analysis of Model $S = \hat{S}eq (1 - e^{-GT})$ for Three Test Distances in Laboratory Systems [47][a]

	Distance		
	9.5 cm	19 cm	38 cm
F	0.29	3.30	0.62
α(F)	>0.75	>0.25	>0.75
$\hat{S}eq$	23.43 ($\pm$5.09)[b]	13.76 ($\pm$3.29)	8.96 ($\pm$1.41)
G	0.17 ($\pm$0.06)	0.32 ($\pm$0.18)	0.32 ($\pm$0.11)
$t_{90\%}$	13.54	7.19	7.19
Species/Day	1.55	1.72	1.12

[a] Lack of fit F and α level attained are presented. $\alpha(F) > 0.01$ is required for decision level.
[b] Asymptotic standard error.

colonization rate was high for the nearest islands. An important point to keep in mind is that the predictions of $\hat{S}eq$ and $t_{90\%}$ both involve extrapolation beyond the actual data. For instance, it can be predicted that the islands at 9.5 cm eventually will equilibrate at 23.43 species ($\hat{S}eq$) and will reach 90% of that number within 13.54 days; yet the actual data show 16 species at the end of 7 days. The fitted model based the greater equilibrium number and longer time to $t_{90\%}$ for the nearest islands on the relatively rapid rate of colonization and the lack of an asymptote up to 7 days, but colonization probably would not continue at this high rate for very much longer. Therefore, caution should be exercised in evaluating these rather preliminary results.

However, the reason that the final time to equilibrium was longer for the nearest islands may be that the high invasion pressure from the epicenter (which still contains many more species) forces a few more species to be added very slowly as new niches become available. (Scanning electron microscopy studies by Paul et al. [22] have shown that the surfaces of the PF units become more complex through time.) When habitat space becomes available on more distant islands (where invasion pressures should be lower), it is more likely to be filled by individuals from expanding populations of species already present on the island than by new invaders. Thus, even though invasion pressure is lower on more distant islands, it is possible that they may reach equilibrium species numbers more quickly than equal size islands nearer the epicenter. Simberloff [2,56] reported that the relatively few species of insects that were early invaders on the most distant mangrove island in his defaunation experiments built up abnormally large population sizes, thus retarding the attainment of the predicted equilibrium number of species.

RESULTS – AREA EFFECT

The accumulation of species with time on islands of different volumes equidistant from epicenters is illustrated in Figure 8. Islands of 26-cm^3 volume accumulated species faster and consistently had slightly higher species richness than larger islands during the 7-day observation period. The noninteractive island colonization model, $S(t) = \hat{S}eq(1 - e^{-GT})$, was fitted to the data, and L.O.F. tests were performed (Table III). The accumulation of species on islands of all three sizes was explained adequately by the island colonization model, showing no significant L.O.F. ($\alpha(F) > 0.01$).

Discussion of Area Effect

The importance of area in determining the number of species present on islands has been well documented in field experiments [57-60]. Experiments by Patrick [32] with freshwater diatoms, Cairns and Ruthven [37] and Kuhn and Plafkin [33] with freshwater protozoa, and Schoener [54] with small marine invertebrates all have shown that increasing size of experimental islands resulted in higher equilibrium species richness.

In our experiments, species richness on the smallest (26 cm^3) islands was consistently higher than on the two larger-volume islands. These results could be viewed as contradictory to the MacArthur-Wilson equilibrium model's prediction that equilibrium species diversity will increase with island size. However, the estimates of $t_{90\%}$ show that not enough time was allowed for all islands to reach equilibrium, so equilibrium species richness on islands could not be compared adequately. The 26-cm^3 islands appear to be close to an equilibrium number of species (17.5) by day 7, with 90% of that number (15.75 species) reached between days 3 and 5. The predictions from the model are an $\hat{S}eq$ of 17.76 with $t_{90\%}$ (15.98 species) by 3.97 days, which are very close to the observed values. With the 207-cm^3 islands, the observed and predicted values for equilibrium species diversity were greatly divergent, but the model did predict that much more time was needed for these islands to approach equilibrium.

The size range between the largest and smallest islands used in these experiments (less than one order of magnitude) may have been too narrow to see a positive correlation between species numbers and volume of islands. In a field study of protozoan colonization where a positive correlation between island size and species number was found

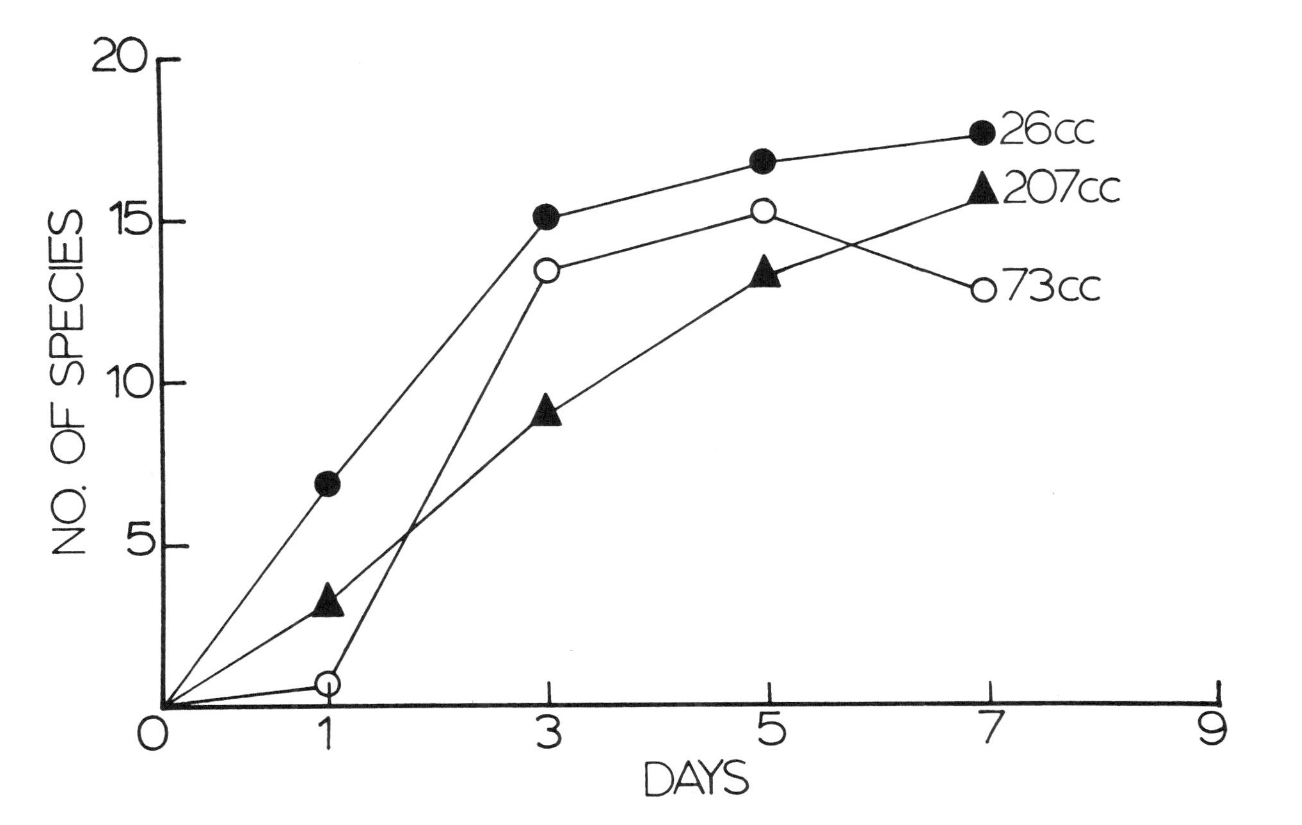

Figure 8. Colonization curves for islands of different volumes located equidistant from epicenters in laboratory microecosystems.

Table III. Nonlinear Regression Analysis of Model $S = \hat{S}eq (1 - e^{-GT})$ for Three Island Sizes (volumes) Used in Laboratory Test Systems[a]

	Island Volume		
	$26\ cm^3$	$73\ cm^3$	$207\ cm^3$
F	0.44	0.83	0.29
α(F)	>0.75	>0.50	>0.75
$\hat{S}eq$	17.76 (±1.62)[b]	15.58 (±6.85)	23.43 (±5.09)
G	0.58 (±0.18)	0.28 (±0.27)	0.17 (±0.06)
$t_{90\%}$	3.97	8.22	13.54
Species/Day	4.03	1.70	1.56

[a]Lack of fit F and α level attained are presented. $\alpha(F) > 0.01$ is required for decision level.
[b]Asymptotic standard error.

[37], the largest islands were 569 times the volume of the smallest and, of the 10 different sizes used, three were smaller than islands used in our microecosystem experiments. It still may be possible to demonstrate the area effect in laboratory systems if a suitable size range of islands is allowed to be colonized for an adequate length of time.

RESULTS – EPICENTER MATURITY

Two distinct patterns of species accumulation can be seen in colonization curves (Figure 9). Faster colonization rates occurred initially on islands exposed to 1- and 3-week age epicenters than those exposed to 1-day and 1-year epicenters. Also, numbers of species on islands tested with 1- and 3-week epicenters reached peaks and then decreased within the experimental period, but species richness was still increasing on islands tested with 1-day and 1-year epicenters.

Table IV presents the results of fitting the noninteractive equilibrium model $S = \hat{S}eq(1 - e^{-GT})$ to the data. All the curves, except for the one for islands exposed to the 1-week epicenter, showed no significant lack of fit, indicating that the island colonization model adequately described the accumulation of species. The predicted $\hat{S}eq$ ranged from 8.73 species for the islands exposed to the 3-week epicenter to 163.03 species for those in the system with the 1-year epicenter.

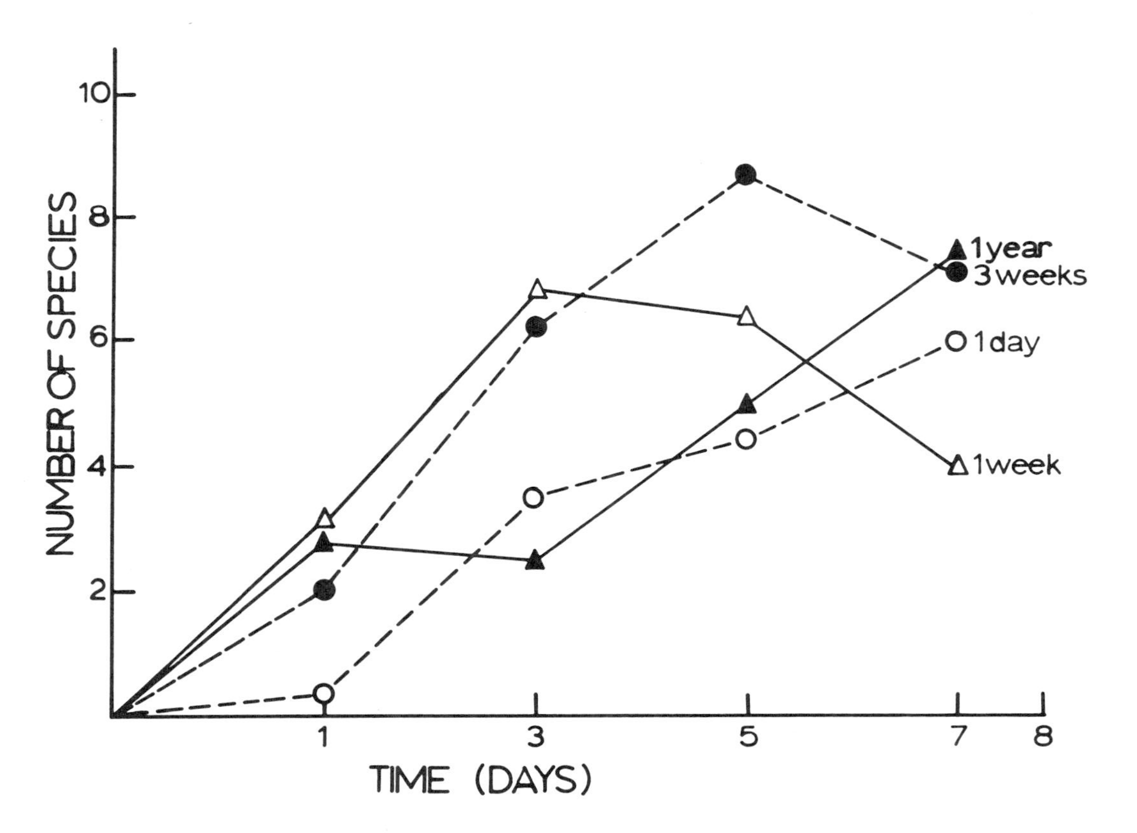

Figure 9. Colonization curves for islands located equidistant from epicenters of different maturities.

Table IV. Nonlinear Regression Analysis of Model $S = \hat{S}eq (1-e^{-GT})$ for Colonization from Epicenters of Various Maturities in Laboratory Systems [47][a]

	Epicenter Maturity (AGE)			
	1 day	1 week	3 weeks	1 year
F	1.88	18.3	0.65	5.53
α(F)	>0.25	>0.001	>0.75	>0.025
$\hat{S}eq$	13.79 (±11.66)[b]	L.O.F.	8.73 (±2.27)	163.03 (±2826.3)
G	0.08 (±0.09)		0.40 (±0.28)	0.006 (±0.096)
$t_{90\%}$	27.58		5.97	352.67

[a] Lack of fit F and α level attained are presented. α(F)>0.01 is required for decision level.
[b] Asymptotic standard error.

Discussion – Epicenter Maturity Experiments

A hypothesis of what may have happened involves consideration of the types and abundances of species on epicenters of various maturities. The 1-day epicenter community consisted almost entirely of r-selected or "pioneer" species. There were small, colorless flagellates such as *Bodo* spp., small green flagellates such as *Chlamydomonas* spp., and small ciliates such as *Cyrtolophosis* sp. These protozoa have been found as early colonists in other studies using PF unit artificial substrates and have been referred to as "pioneer species" [61,62]. The "pioneer"-type protozoa also have been found to have high dispersal or vagility indices [55]. However, because this epicenter was only in the lake for 1 day, the abundance of each species was relatively low, so invasion pressure on surrounding islands probably would be very low. The 1- and 3-week epicenters (particularly the 1-week) still had relatively high percentages of pioneer species, and the abundance of each species generally was higher than in the 1-day epicenter. This could have resulted in relatively high invasion pressures on islands and may have caused the fast immigration rates and high species diversity within the 7-day experimental periods. The 1-year epicenter had the highest species richness at initial examination; therefore, it might be expected that islands exposed to it would rapidly reach a high diversity. However, the 1-year epicenter community had a relatively low proportion of r-selected, pioneer-type species. The majority of its species were larger and more complex ciliates, which seem adapted for competition for food and habitat space in mature communities. These protozoa, which include spirotrichs, hypotrichs and peritrichs, would not be suited for rapid colonization of barren islands.

Again, there was a problem because of running the experiments for only 7 days. The model predicted that islands exposed to 1-day epicenter would reach an $\hat{S}eq$ of 13.79 in 27.58 days, which seems nearly reasonable (the 1-day epicenter had 13 species initially, and there is reason to believe that most could reach the islands), but the $\hat{S}eq$ of 163.03 in 352.67 days for islands tested with 1-year epicenters is surely an artifact of not allowing sufficient time for colonization! The model cannot extrapolate so far beyond the data with any degree of accuracy.

THE EFFECT OF EPICENTER MATURITY ON THE PROCESS OF ISLAND COLONIZATION– THE ROLE OF PIONEER SPECIES

As an extension of our previous microecosystem experiments, we studied more intensively the role of pioneer species in the process of island colonization. We wanted to follow the colonization process until relatively stable equilibria were attained on PF unit islands in the microecosystems and develop a system for designating particular species of protozoa as "pioneer" species, as such evidence became available. The hypothesis was that islands exposed to epicenters, which themselves were at early stages of the colonization process, would be colonized at a faster rate than islands exposed to mature communities (those at or near equilibrium) because of a greater proportion of pioneer species in the developing communities. The influence of the species composition of an epicenter community on the colonization of nearby islands has not heretofore been tested as rigorously as the effects of island size and distance from an epicenter on colonization rate. Yet a consideration of the species composition of an epicenter may be as important as distance from the epicenter and size of the target area in predicting the rate of colonization of an island or the rate of recovery of a damaged ecosystem.

Materials and Methods

The apparatus used in these experiments was identical to those used in earlier experiments [47], but the procedure was modified so that each island was harvested only once. Four replicate island PF units were harvested from each system on days 1, 3, 5, 7, 9, 11, 13 and 21 after setup (no 21-day harvest for island tested with the 3-week source pools). Source pool and island PF units were positioned as shown in Figure 10.

Results

Species numbers on source pools increased in proportion to the length of time allowed for their colonization in Pandapas Pond (Table V). The percentage of species surviving on the source pools after 21 days in the colonization trays varied from greater than 100% for one of the 3-day sources to a low of 38% for one of the 3-week source pools. In determining the percentage of species reaching islands, a species was counted if it was found on any island at any time during a particular experiment. The 3-day source pools contributed the highest percentage (96%) of their species to the surrounding islands, while only about 40% of species from the 13-week source pools reached islands in the test systems.

Figure 11 illustrates the accumulation of protozoan species with time on islands exposed to source pools of different maturities (each point on the graph is the mean of four observations). Those exposed to source pools of 3 days and 1 week maturity reached peaks in species numbers by day 5. Peaks in species richness on islands tested with 3- and 13-week source pools occurred on days 9 and 11, respectively. After these peaks, dynamic equilibria appeared to develop in all systems.

The MacArthur-Wilson noninteractive colonization model, $S(t) = \hat{S}eq(1 - e^{-GT})$, was fitted to the experimental data. Islands tested with 3-day source pools were estimated to equilibrate the most rapidly ($G = 0.323$, $t_{90\%} = 7.12$ days), with islands exposed to source pools of greater maturity taking progressively longer to reach equilibrium numbers of species (Table VI).

Discussion

During the 21 days allowed for colonization in the laboratory experiments, the 3-day colonized source communities contributed 100% of their species numbers to islands in the test systems. The more mature source communities contributed only 39.5% (mean for 13-week colonized sources) or 63% (mean for 3-week sources) of their species to the experimental islands. It may be that the 3-day colonized source communities had higher values as epicenters than the more mature communities because they contained relatively more pioneer species.

Pioneer species generally have r-selection strategies. These strategies

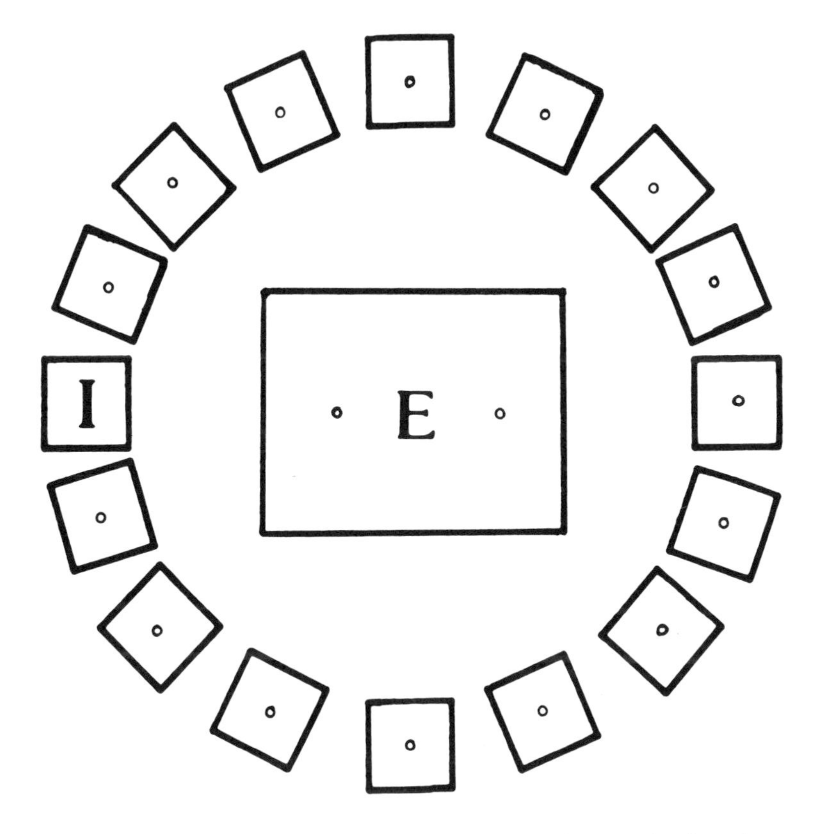

Figure 10. Placement of island and epicenter substrates in the microecosystems. Distance between epicenter (E) and island (I) PF units is approximately 9.5 cm.

Table V. Numbers of Protozoan Species on Source Pool Communities used in Laboratory Microecosystem Experiments [63]

	Source Pool Colonization Time							
	3 days		1 week		3 weeks		13 weeks	
	A	B	A	B	A	B	A	B
No. Species at Start	13	10	31	— [a]	37	35	54	57
No. Species at Finish	13	11	15	—	14	15	26	27
Species Survival (%)	100	110	48	—	38	43	48	47
Species Reaching Islands (%)	92	100	54	—	52	74	44	35

[a] No "B" series with 1-week source pool.

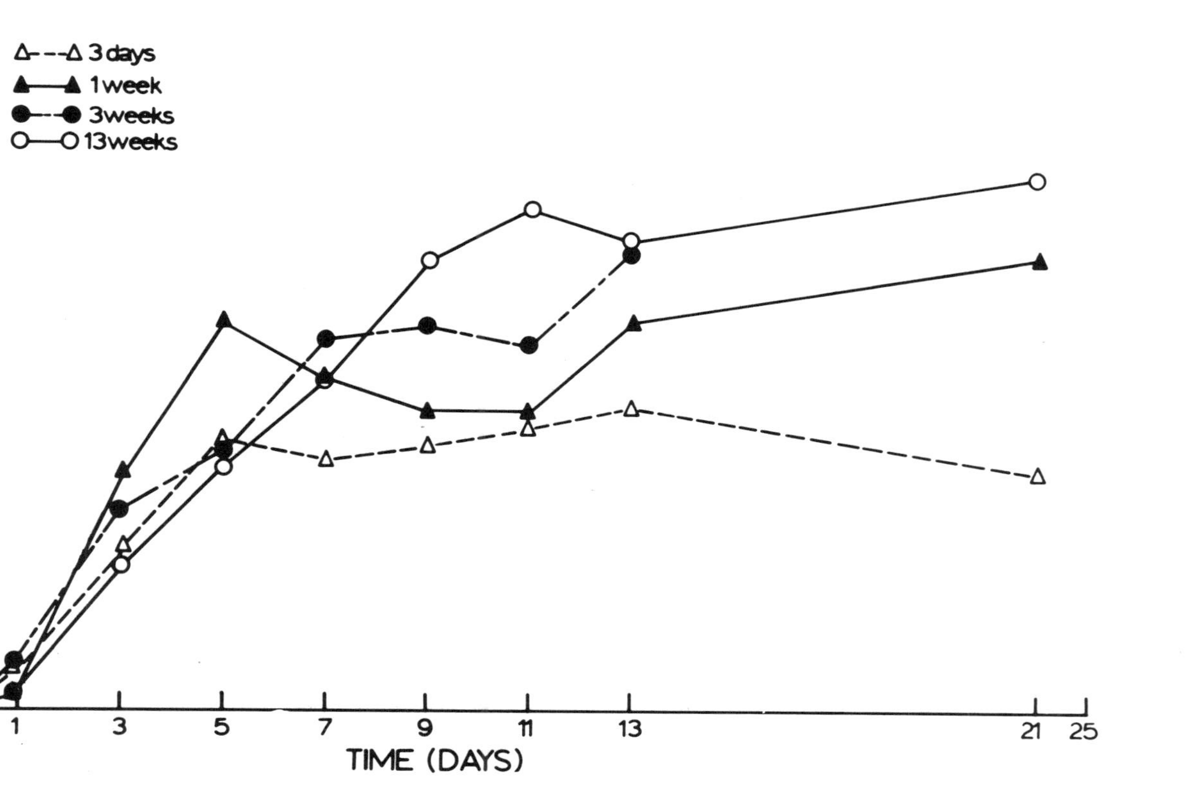

Figure 11. Colonization curves for islands exposed to source pools of different maturities.

Table VI. Nonlinear Regression Analysis of Model $S = \hat{S}eq(1-e^{-GT})$ for Islands Exposed to Sources of Different Maturities [63][a]

	Source Pool Maturity			
	3 days	1 week	3 weeks	13 weeks
F	2.77	8.88[3.33][b]	0.54	3.59
α(F)	0.025	>0.001[>0.01]	>0.75	>0.01
$\hat{S}$eq	10.77 U=11.69[c] L= 9.85	L.O.F.[15.27] U=17.23 L=12.62	17.87 U=22.15 L=13.60	23.83 U=28.72 L=18.94
G	0.323 U=0.421 L=0.225	[0.238] U=0.375 L=0.101	0.171 U=0.249 L=0.093	0.105 U=0.144 L=0.066
$t_{90\%}$	7.12	[9.6]	13.46	21.92

[a] Lack of fit F and α level attained are presented. $\alpha(F) > 0.01$ is required for decision level.
[b] Values without data for days 7, 9 and 11.
[c] Upper and lower asymptotic 95% confidence limits.

include high reproductive rates, density-independent mortality, and the ability to utilize variable or unpredictable resources in unsaturated, nonequilibrium communities [64]. Another characteristic of pioneer species is their high degree of dispersibility, which is an adaptation to habitats that are temporary in space and time [65].

Little information is available on the autecology of protozoa, so it was necessary to adopt an operational definition of pioneer species based on data from previous colonization studies. A full description of the methods used to produce a list of pioneer protozoan species is given by Henebry and Cairns [63]; the list included mostly protozoa that colonized PF unit islands by day 3 in previous studies or had a high dispersal index in Maguire's studies [55].

Table VII summarizes the occurrence of these pioneers in epicenters of different maturities used in our experiments. Of species seen in the 3-day colonized source communities, 93% were pioneers. For the other source communities the percentage of pioneer species ranged from 13% to 29%.

The very high percentage of pioneer species in the 3-day epicenter communities probably was a major factor in the rapid colonization of islands exposed to them. The islands exposed to 3-day and 1-week epicenters had high colonization rates (G-values) and quickly reached equilibrium species numbers (Figure 11). In the former case, this may reflect mostly the dispersal capacity of the pioneer species that comprised nearly the entire species pool in the 3-day epicenter community. The 1-week epicenter did have the second highest proportion of pioneers (29%); however, more

Table VII. Occurrence of Protozoan Species with High Pioneer Ratings on Source Pools Used in Microecosystem Experiments [63]

Species	Pioneer Rating	High Pioneer Value	Age of Source Pool			
			3 days	1 week	3 weeks	13 weeks
Monas sp.	4	x	x	x	x	x
Cyathomonas truncata	4	x	x	x	x	x
Chlamydomonas sp.	5	x	x	x		
Peranema inflexum	3	x	x		x	x
Ochromonas sp.	2	x	x			
Chilomonas paramecium	2	x	x			
Anisonema pusillum	2	x	x	x		
Cryptomonas erosa	3	x	x	x		x
Urotricha agilis	2	x	x	x	x	x
Cinetochilum margaritaceum	2	x	x	x		x
Glaucoma scintillans	2	x	x	x	x	
Cyclidium musicola	3	x	x	x	x	
Cyclidium litomesum	3	x	x			x
Hemiophrys sp.	1		x			
Mean No. of Species on Source Pools			14	31	36	55.5
No. Species with High Pioneer Value			13	9	6	7
Percentage of Species with High Pioneer Value			93	29	17	13

importantly, pioneer species may have made up a much larger proportion of the total organisms than the 29% figure would indicate. The abundance of most species was higher in the sample from the 1-week epicenter community than in the sample from the source exposed for only 3 days. The high proportion of pioneer species in their source communities may explain the high G-values (a reflection of colonization rate) of the colonization curve for islands exposed to the 3-day and 1-week source communities.

Patrick [32] has shown that source pool (epicenter) species diversity can affect the number of diatoms found at equilibrium on experimentally placed glass slide islands. She found that species numbers on the slides equilibrated at 14–29 species with a source pool of 60 species and at 160 species when exposed to a pool of 250 species. In our experiments, islands exposed to increasingly mature source communities (with increasingly greater species diversities) accumulated significantly greater numbers of species.

In our experiments, the colonization process was followed through time. This allowed observation of the effect of epicenter community size and composition on colonization rate, as well as the effect of source pool size on ultimate species richness on islands. The parameters ($t_{90}\%$, G and $\hat{S}eq$) estimated as a result of fitting the noninteractive model $S(t) = \hat{S}eq(1 - e^{-GT})$ lent support to the hypothesis that islands exposed to epicenter communities of increasing maturity (increasing species diversity) take progressively longer to reach eventually higher equilibrium species numbers. These estimates also lend support to a similar hypothesis proposed by Cairns et al. [49] as a reason to expect the attainment of equilibrium number of species to be more rapid for PF unit islands drawing colonists from a stressed species pool in a lake situation. They suggested that a stressed community is commonly dominated by r-selected opportunistic species (analogous to pioneer species in our laboratory experiments), which are well adapted to the pioneer episodes of colonization.

Further, barren islands drawing propagules from this type of pool would be expected to accumulate species more rapidly than islands drawing from a more complex source (such as the more mature source communities in our laboratory experiments). The results also supported the hypothesis that colonization rates on islands were influenced by the proportion of pioneer species in source communities. The proportion of pioneer species decreased with the increasing maturity of source communities. Finally, controlled laboratory experiments extended and clarified the results of field work on the colonization process by Patrick [32], Opler et al. [66] and Cairns et al. [49], and quantitatively confirmed that kinds as well as numbers of species in epicenters could be very important in determining the rate of colonization of islands. Species composition of epicenters also may be of as much or greater importance than size or distance of a nearby source pool in determining the rate of recovery of a damaged ecosystem.

COLONIZATION OF PF UNIT SUBSTRATES BY PROTOZOA IN A RIVER

Polyurethane foam units were used as artificial substrates to sample the protozoan populations at each of the five stations in the South River, Virginia. Fifteen PF units measuring 76 mm × 64 mm × 30 mm were placed at each station. The units were tied to a line that was anchored to a cinderblock and positioned such that they floated just below the water surface. Water depth at each station was approximately 0.28–0.30 meters. Units were sampled 0.5 day, 1 day, 2 days, 4 days and 7 days after placement.

Results

Table VIII lists the numbers of protozoan species on each PF unit collected from the river. Units from stations 3 and 4, just downstream of a municipal sewage effluent, generally had greater species richness than PF units from stations 1, 2 and 5. Colonization curves for two extreme stations, station 1 (least species richness) and station 4 (greatest richness), and the average of all stations graphically illustrate that there was essentially no further species accumulation after day 1 (Figure 12).

Discussion

It appears that equilibrium species numbers were reached extremely rapidly on the PF unit islands. In studies of over 40 lakes in Virginia, Michigan and Colorado, it has been found that 2–5 weeks were required for attainment of equilibrium numbers of species on the PF units. In both lotic and lentic waters, the method of protozoan dispersal to the PF units was assumed to be mainly by water currents, and the major source (or epicenter) was assumed to be the benthos. In the South River, naturally there was much greater current than in any lakes we have investigated. Also, the PF units were located much closer to the benthos (0.28–0.30 meter in the river vs 2–3 meters in the lakes). These two factors alone may account for the great increase in colonization rates of PF units located in the river. Although our experience with PF units in lotic waters is limited, it seems that 3–7 days of immersion in a river allows enough time to establish an equilibrium on PF unit substrates. Within 1–2 days, following a decline in number, a maximum number of protozoan species was observed in PF unit samples from a cave stream [68] and from Crab Creek, Montgomery County, Virginia [69].

COLONIZATION OF PF UNIT SUBSTRATES IN WETLANDS

PF units serving as islands for protozoa in some wetland lakes in northern Lower Michigan [70] were colonized as rapidly as PF unit islands in rivers (Figure 13). When the noninteractive model was fitted to the data, the predicted $t_{90\%}$ for PF unit islands from both bog lakes and one marsh site was less than 3 days (Table IX). Islands at the fen sites appeared to colonize the most rapidly; over 90% of maximum numbers at species were found after one day of exposure, but the noninteractive model did not fit their data.

Table VIII. Number of Species on PF Units Sampled from the South River, Waynesboro, Virginia [28]

	Station									
	1		2		3		4		5	
Sample Day	A	B	A	B	A	B	A	B	A	B
0.5	19	23	23	23	22	30	29	36	37	35
1	28	36	35	40	43	41	51	52	35	38
2	31	34	31	39	46	47	47	45	38	42
4	25	32	32	39	49	50	49	54	41	45
7	35	33	24	28	52	52	37	34	40	42

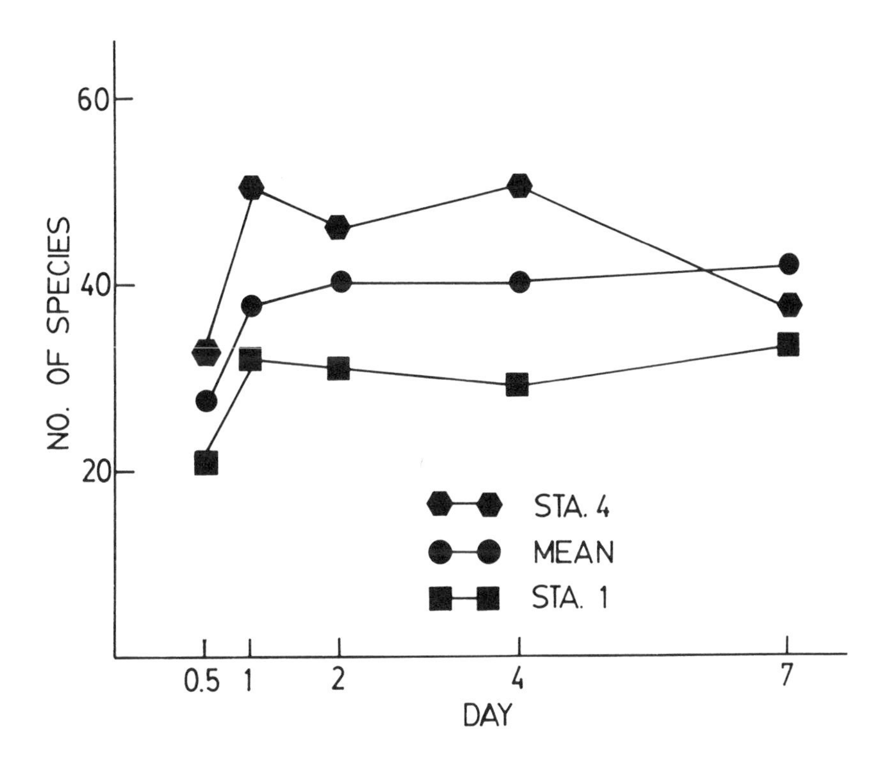

Figure 12. Species accumulation on PF unit substrates at Stations 1 and 4, and the average number of species accumulated at all stations in the South River [67].

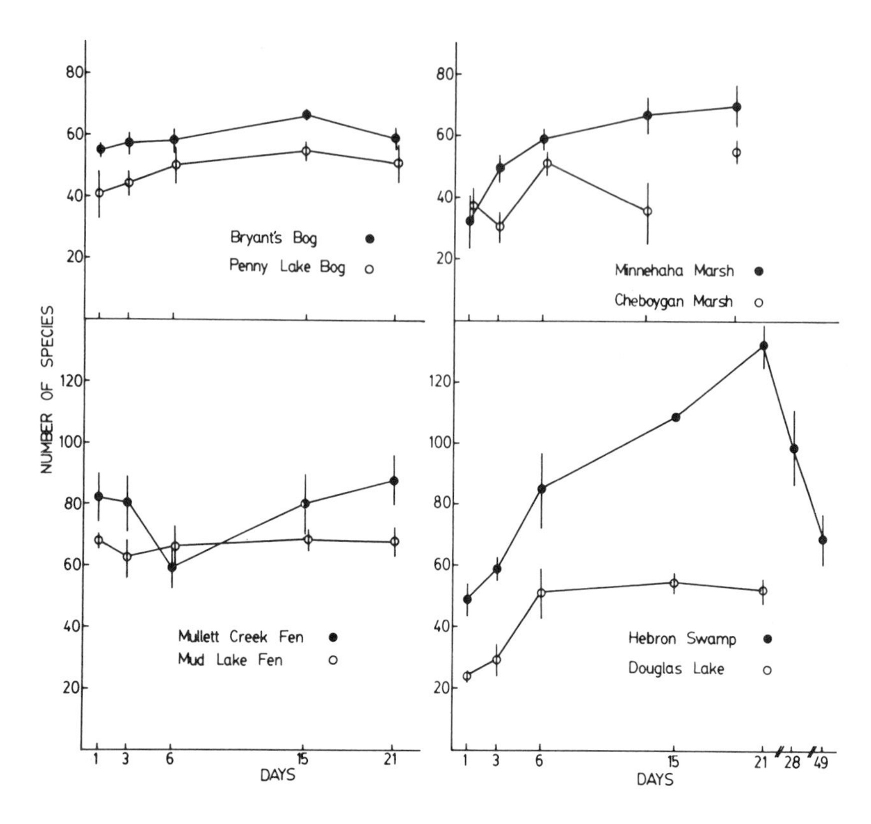

Figure 13. Protozoan colonization curves for wetlands sampled in the summer of 1977. Each point represents the mean of four PF unit substrates; bars represent one standard error of the mean.

Several theories could explain the extremely rapid colonization at wetland sites. Cairns et al. [49] have related the rate at which PF units were colonized by protozoa to the degree of eutrophication of lake ecosystems. In their study of Smith Mountain Lake, Virginia, an impoundment with a well-documented trophic gradient from eutrophic upriver near a city to oligotrophic at the site of the dam, they found that colonization rates, as measured by G-values, were much higher at the eutrophic end. The G-values at both Michigan bog sites and the marsh were higher than those reported for the eutrophic end of the Smith Mountain Lake. Subsequent studies of primary productivity of phytoplankton and nutrient status in these and other wetland sites [71] have shown that bog and fen lakes can be highly productive. Cairns et al. [49] presented evidence that r-selected protozoa predominated at eutrophic stations in Smith Moun-

Table IX. Nonlinear Regression Analysis of Model S = $\hat{S}eq$ $(1-e^{-GT})$ for Colonization of PFU Substrates in 11 Aquatic Systems

Site	Year Sampled	G	$\hat{S}eq$	$t_{90\%}$
Bryant's Bog	1977	2.42±0.56	60.26±1.32	1.73
Penny Lake Bog	1977	1.48±0.38	51.38±2.07	2.83
Mud Lake Fen	1977	L.O.F.	L.O.F.	L.O.F.
Mullett Creek Fen	1977	L.O.F.	L.O.F.	L.O.F.
Minnehaha Marsh	1977	0.52±0.07	66.32±2.42	8.06
Cheboygan Marsh	1977	1.64±0.44	46.25±4.64	2.56
Hebron "Swamp"	1977	0.34±0.11	100.16±7.61	12.32
Douglas Lake	1977	0.37±0.06	53.57±2.49	11.33
Walloon Lake	1978	0.13±0.03	46.74±3.58	32.24
Dog Lake	1978	0.21±0.76	37.72±3.71	19.96
Munroe Lake	1978	0.91±0.86	53.18±7.07	4.60

tain Lake and hypothesized that species pools dominated by r-selected species tended to furnish colonists to the PF units at a faster rate than other types of species pools.

Another explanation we considered for the very high colonization rates in some wetlands was that PF units at some densely vegetated sites were in contact with emergent and floating vegetation. The PF units may have acted more as habitat patches [2,72] rather than mimicking oceanic islands. We have largely abandoned this hypothesis since PF units were in close contact with emergent vegetation in all the wetland lakes, yet colonization rates varied greatly. Also, we tried locating PF units at various distances from the *Sphagnum* spp. mat (the assumed epicenter for colonists) in two bogs and found no significant difference in colonization rates [71].

LONG-TERM COLONIZATION – EVIDENCE FOR A SECONDARY EQUILIBRIUM STAGE IN ISLAND COLONIZATION

Cairns and other investigators from Virginia Polytechnic Institute and State University have used PF units to study colonization rates in Douglas Lake, Michigan, every summer since 1968. Equilibrium species numbers of protozoa have been reached on the PF units within 2–5 weeks, generally within 2–3 weeks. The predicted equilibrium numbers, $\hat{S}eq$, G-values and $t_{90\%}$ are given for this period where the noninteractive model fit the data (Table X). Predicted $\hat{S}eq$ and $t_{90}\%$ ranged from 39.7 to 53.6 species and 11.3 to 38.4 days, respectively. Yongue and Cairns [24] found that the number of species found on PF units that had remained in Douglas Lake and other lakes for one year were essentially the same as the equilibrium number of species on units sampled the previous summer.

In the summer of 1980 we had the opportunity to sample PF units that had been in place in Douglas Lake for 3 and 5 years. These units had been extras from studies carried out in summers of 1975 and 1977. Although only two units were examined from each of these years, it was readily apparent that the number of species on these old units were substantially lower than numbers observed on units placed there in the summer of 1980 (Table XI) and about 10 species lower than the $\hat{S}eq$ predicted from fitting the noninteractive model to the colonization data collected in 1980. It occurred to us that possibly the number of species of protozoa in Douglas Lake were unusually low in the years when the older units were initially colonized. However, the number of species observed on PF units after 21 days in the lake in 1975 and 1977, the years when the 3- and 5-year-old units were initially colonized, were within the "normal" range observed in Douglas Lake (Table XI) and were within the normal range for predicted $\hat{S}eq$ (Table X). Thus, the lower numbers observed after 3–5 years did not seem to be a consequence of the year in which the PF units initially were colonized. Data collected in 1974 in which some 1- and 2-year-old PF units were censused, along with units that had been placed that summer, showed that the number of species remained about

Table X. Results of Nonlinear Regression Analysis of Model $S = \hat{S}eq\,(1-e^{-GT})$ on Colonization Data from Douglas Lake, Michigan (1974–1980)

Year	G-value	$\hat{S}eq$	$t_{90}\%$ (days)
1974	L.O.F.	–	–
1975	0.258±0.049	42.10±2.58	16.24
1976	0.246±0.097	39.69±4.95	17.03
	0.236±0.087	41.43±4.86	17.75
1977	0.370±0.060	53.57±2.49	11.33
1978	L.O.F.	–	–
1979	L.O.F.	–	–
1980	0.120±0.037	42.90±3.64	34.92

Table XI. Numbers of Species of Protozoa on Polyurethane Foam Substrates from Douglas Lake, Michigan

	Length of Exposure of PFU Substrates												
	7 days	15 days	21 days	21 days	21 days	21 days	28 days	41 days	50 days	1 yr	2 yr	3 yr	5 yr
Year Sampled	1980	1980	1974	1975	1978	1980	1980	1980	1980	1974	1974	1980	1980
No. Replicates	3	3	10	8	3	3	3	3	3	10	10	2	2
No. Species	24.0	34.0	44.8	37.9	43.3	37.7	39.3	49.7	39.7	38.2	28.7	29.0	27.5

the same after 1 year in the lake but decreased by about 10 species after 2 years (Table XI).

The wide variation in colonization rates and equilibrium numbers over the years is probably a consequence of conditions existing at the time the PF units were placed in the lake. Cairns et al. [73] have described how location in the lake may affect equilibrium numbers, and Cairns et al. [27] have shown that seasonal variation, particularly lake overturn, affects colonization rates and equilibrium numbers on PF units. Schoener [54] and Osman [72,74] also have found that seasonality can affect colonization rate and equilibrium numbers of species in marine invertebrate island communities. Presumably, the chemical and physical conditions existing in the lake in the particular year of PF unit placement (regardless of season) also cause variations in equilibrium numbers. Therefore, it is all the more remarkable that the 2- to 5-year-old PF unit communities remained very stable with regard to numbers of species. It is probable that some internal stabilizing mechanism is operative and some basic changes have occurred in communities that have been allowed to form on PF units for 2 or more years.

Figure 14 presents a model of what may occur during the colonization of PF unit islands by protozoa. The model divides the processes involved in establishing and maintaining a particular diversity on the PF unit islands into three phases: (1) noninteractive phase; (2) early interactive phase; and (3) secondary interactive phase. The predictions of the MacArthur-Wilson [35,51] species equilibrium model seem to describe only the noninteractive phase, characterized primarily by colonization, and part of the early interactive phase, characterized by the establishment of a dynamic equilibrium. As noted by Osman [72], the species equilibrium theory is an extremely robust model for interpreting diversity. After the period of initial colonization, however, it seems that factors other than immigration and extinction rates may become more important in maintaining diversity on islands. Osman [72] has shown that seasonality and the natural instability of his rock islands affect equilibrium numbers of species; again however, his model is concerned with changes in immigration and extinction rates, and thus would be limited to the noninteractive and early interactive phase of our model. Many investigators have speculated on how other factors, particularly competition among members of the island community, may be involved in maintaining species diversity at a particular level, but so far these competitive interactions have not been demonstrated experimentally [75,76].

Simberloff [56] pointed out that new islands are likely to reach a short-lived noninteractive equilibrium and then rather quickly shift to an interactive equilibrium as populations of individual species increase.

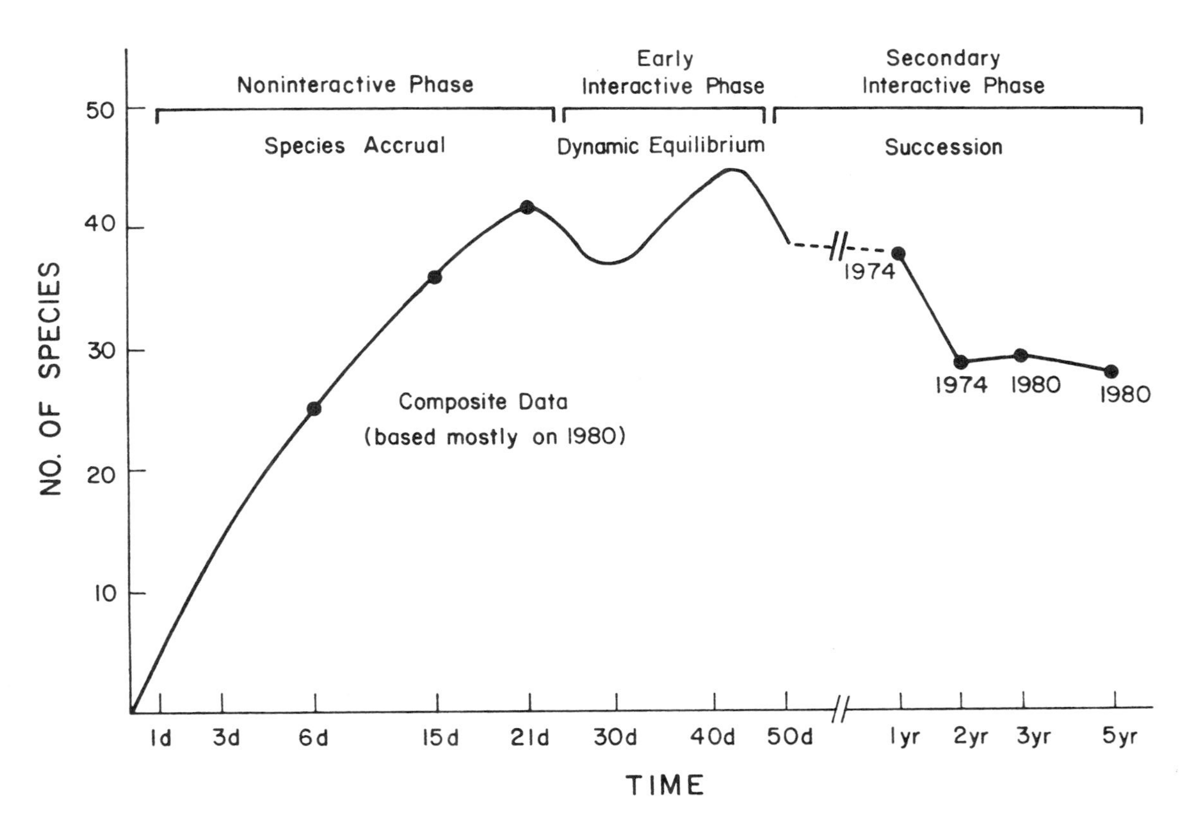

Figure 14. Long-term colonization of PF unit artifcial islands by protozoa.

Wilson [77] said the interactive equilibrium could be higher than the noninteractive if synergistic effects prevail, or lower than the noninteractive if interactive extinctions of species increases. Simberloff and Wilson [78] speculated that following the initial noninteractive colonization process there would be a phase termed the "interactive" equilibrium. This would be succeeded by an "assortive" equilibrium that would be higher because of combinations of species that are either better adapated to the conditions of the local environment or else were able to coexist longer with the particular set of species among whom they find themselves. Simberloff [2] expanded this concept of multiple equilibria to include evolutionary and taxon cycle equilibria. His article predicted that noninteractive and interactive (including assortive) would be achieved on an ecological time scale, and evolutionary and taxon cycle equilibria would occur on an evolutionary time scale. Each equilibrium would be a balance of forces tending to increase or decrease the numbers of species on an island.

Our model for the long-term "colonization" process in protozoan communities (Figure 14) probably extends through at least the assortive and interactive equilibria in Simberloff's [2] scheme and possibly into the evolutionary equilibrium stage after 2–5 years. The terms *species accrual, dynamic equilibrium* and *succession* refer to the dominant process occurring in each of the three phases of our model. Use of the term *colonization,* the process of species accrual over time, perhaps should be restricted to events in the noninteractive phase; however, since there is precedent for calling the overall process colonization, we will call the process in the noninteractive phase *species accrual.* We have much experimental evidence, as previously cited, that species accural is the dominant process on the PF unit islands up to about 2–3 weeks in protozoan communities. Cairns et al. [52] have shown that the process of species accrual can be explained by changes in immigration and extinction rates through time. The noninteractive phase often ends in a peak of species abundance before the effects of interaction (presumably competition between species for resources and/or habitat space) become important. There is also a fairly large body of evidence for the early interactive phase of island colonization where species accrual becomes less important and interactions between species are thought to increase [25,29,34,61,72,73,78,79]. The dominant process here is the dynamic equilibrium in which a sorting of species occurs because some early colonists would be less adapted to exist on the island than some later arrivals. There would be lowering of equilibrium numbers as some of these early colonists became extinct on the island and higher equilibria as coadapted subsets of species increased.

As discussed previously, we have limited experimental evidence for a third major phase in the colonization process in protozoan communities. We have chosen to call this the secondary interactive phase mainly because there is little direct evidence as to what forces are acting to produce this equilibrium. This phase may correspond to the evolutionary equilibrium in the Simberloff [2] scheme in which the main force is increased extinction as niches evolve to become narrower, but we have no direct evidence for this. However, we do have some evidence that a dominant process of this phase is succession. Immigration and extinction and interspecific interactions would still occur, but probably are not as important in this phase of the colonization process. We also have some evidence that silt and debris accumulate in the PF units over a period of time (calcium carbonate deposits also occur after one year), which decrease available habitat for protozoans. Still, the remarkable consistency in numbers and kinds of species in 2- to 5-year-old PF units indicates the presence of fundamental changes in the biotic community.

One change in the PF unit island communities that seems to be general is a decrease in the percentage of flagellates with increasing age of the protozoan communities. Yongue and Cairns [62] found that flagellates reached equilibrium much earlier than other taxonomic groups, then declined in importance as more species of ciliates and sarcodines colonized the PF unit islands. The trend of decreasing numbers of flagellates seemed to hold in our long-term studies of colonization (Table XII). The flagellate:ciliate (F:C) ratio steadily decreased from about 1.0 in 21-day-old island samples from several different years to a low of 0.25 in three-year-old PF unit island samples. The F:C ratio increased slightly in the five-year-old samples, but this may have been an artifact of the small sample size; the general pattern is one of a decreasing F:C ratio as the islands mature. The F:C ratio roughly corresponds to the autotroph to heterotroph ratio in protozoan communities. A decrease in the autotroph to heterotroph ratio, a structural attribute, may indicate succession in a community much as a decrease in the P:R ratio, a functional attribute, indicates succession in an ecosystem. The P:R ratio (ratio of gross production to community respiration) is typically greater than one in early stages of ecological succession, and the P:R ratio then decreases to less than one as the ecosystem matures [45,80,81].

Thus, it appears that we may be observing succession during the secondary interactive phase of colonization. Succession is generally assumed to result in an increase in stability of a community, although many would challenge this assumption [82]. From our admittedly limited data, it does appear that older PF unit island protozoan communities, which seem to have undergone succession in the direction of a decreased autotroph to heterotroph ratio, are more resistant to whatever yearly

Table XII. Flagellate:Ciliate Ratios on Polyurethane Foam Artificial Substrates Colonized for 21 Days to 5 Years in Douglas Lake, Michigan

	Length of Exposure of PFU Substrates									
	21 days	**21 days**	**21 days**	**28 days**	**41 days**	**50 days**	**1 year**	**2 years**	**3 years**	**5 years**
Year Sampled	1975	1978	1980	1980	1980	1980	1974	1974	1980	1980
No. Replicates	8	4	3	3	3	3	10	10	2	2
Flagellate:Ciliate Ratio – $\bar{X}$	0.95	0.98	1.02	0.71	0.60	0.62	0.43	0.36	0.25	0.42

cycles cause a fairly wide fluctuation in numbers of species on recently colonized PF unit islands in Douglas Lake.

CONCLUSIONS

Various aspects of the MacArthur-Wilson noninteractive model were tested using protozoan communities on artificial islands in both field and laboratory experiments. Laboratory experiments produced results comparable to those obtained in the field. In addition, the effects of epicenter (source pool) maturity and the distance of islands from the epicenter were easily tested in the laboratory systems; these would be very difficult to test in field experiments involving microbial communities. The size of an island, its distance from an epicenter, its maturity, the species composition of the epicenter community, the season during which the islands were colonized and the quality of the surrounding water (particularly its trophic status) all seemed to influence the colonization rate and equilibrium number of species on islands. These factors seem to exert their greatest effect during the noninteractive phase of colonization, which is characterized by species accumulation and a temporary equilibrium and which, in turn, is produced by a balance of immigration and extinction of colonists.

We have attempted to delineate events beyond the noninteractive phase of island colonization based on the long-term colonization of PF units by protozoa. Following the noninteractive phase, there is an "early interactive phase" characterized by a dynamic equilibrium that still seems to be influenced considerably by factors external to the island community. The varied response of different species in the island community to lake overturn, trophic state of the lake and maturity of the epicenter community are some factors that seem to be important in producing the dynamic equilibrium in the early interactive phase. Schoener [54] and Osman [74] also have discussed the effects of seasonality in this phase of the colonization process.

We have the least evidence for the "secondary interactive phase" of our model, which we speculate is influenced less by forces external to the island community (i.e., immigration, seasonal changes) but dominated instead by interspecific associations and community succession. Attempts to demonstrate the influence of interspecific associations on species equilibria in island communities have been largely unsuccessful. However, we have found evidence for a type of succession in protozoan communities involving a decrease in the ratios of autotrophic to heterotrophic species. Also, after two years, a secondary number of species, somewhat lower than the dynamic equilibrium, seems to become established on the PF unit islands. Numbers of species seem to remain very stable after about two years of colonization. This "secondary interactive phase" may represent an early phase of the evolutionary equilibrium postulated by Simberloff [2]. Our model is based on very limited evidence; however, we have presented it here in hopes of stimulating further interest in the process of island colonization beyond the initial noninteractive phase.

ACKNOWLEDGMENT

The basic support for this research was provided by U.S. Department of Energy grant EY-76-S-05-4939.

REFERENCES

1. Picken, L. E. R. "The Structure of Some Protozoan Communities," *J. Ecol.* 25(2):324–368 (1937).
2. Simberloff, D. S. "Equilibrium Theory of Island Biogeography and Ecology," *Ann. Rev. Ecol. Sys.* 5:161–182 (1974).
3. Cairns, J., Jr., and W. H. Yongue, Jr. "Factors Affecting the Number of Species in Freshwater Protozoan Communities," in *Aquatic Microbial Communities,* J. Cairns, Jr., Ed. (New York: Garland Publishing, Inc., 1977), p. 257.
4. Cairns, J., Jr., G. R. Lanza and B. C. Parker. "Pollution Related Structural and Functional Changes in Aquatic Community with Emphasis on Freshwater Algae and Protozoa," *Proc. Acad. Nat. Sci. Phil.* 124(5):79–127 (1972).
5. Preston, F. W. "The Commonness, and Rarity of Species," *Ecology* 29:254–283 (1948).
6. Patrick, R., M. H. Hohn and J. H. Wallace. "A New Method for Determining the Pattern of the Diatom Flora," *Notulae Naturae, Acad. Nat. Sci. Phil.* 259:1–12 (1954).
7. Cairns, J., Jr. "Probable Existence of Synergistic Interactions Among Different Species of Protozoans," *Rev. Biol.* 6(1–2):103–108 (1967).
8. Cairns, J., Jr. "Protozoa," in *The Catherwood Foundation Peruvian-Amazon Expedition* (Philadelphia: Academy of Natural Sciences, 1966), p. 53.
9. Bovee, E. C. "Protozoa of Amazonia and Andean Waters of Colombia, South America," *J. Protozool.* 4:63–66 (1957).
10. Kahl, A. "Wimpertiere oder Ciliata (Infusoria)," in *Die Tierwelt Deutschland Urtiere oder Protozoa,* F. Dahl, Ed. (Jena: I. J. Fisher, 1935).
11. Patrick, R. "A Study of the Number and Kinds of Species Found in Rivers in the Eastern United States," *Proc. Acad. Nat. Sci. Phil.* 113(10):215–258 (1961).

12. Robertson, T. B. "The Influence of Mutual Contiguity upon Reproduction Rate and the Part Played Therein by the <<X-Factors>> in Bacterized Infusions Which Stimulate the Multiplication of Infusonia," *Biochem. J.* 15:1240–1247 (1921).
13. Robertson, T. B. "Allelocatalytic Effects in Cultures of *Colpidium* in Hay-Infusion and in Synthetic Media," *Biochem. J.* 18:612–619 (1924).
14. Nalewajko, C. "Extracellular Release in Freshwater Algae and Bacteria: Extracellular Products of Algae as a Source of Carbon for Heterotrophs," in *Aquatic Microbial Communities,* J. Cairns, Jr., Ed. (New York: Garland Press, Inc., 1977), p. 589.
15. Lilly, D. M., and R. H. Stillwell. "Probiotics: Growth Promotion Factors Produced by Microorganisms," *Science* 147:747–748 (1965).
16. Stillwell, R. H. "A Stimulatory Effect on Growth of *Paramecium caudatum* by Products of *Colpidium campylum,*" M. S. Thesis, St. John's University, Jamacia, NY (1962).
17. Cairns, J., Jr. "The Environmental Requirement of Freshwater Protozoa," in *Biological Problems in Water Pollution Third Seminar,* Robert A. Taft Sanitary Engineering Center, Cincinnati, Ohio, PHS Publication No. 999-WP-25 (1962), p. 48.
18. Patrick, R. "A Proposed Biological Measure of Stream Conditions Based on a Survey of the Conestoga Basin, Lancaster County, Pennsylvania," *Proc. Acad. Nat. Sci. Phil.* 101:277–341 (1949).
19. Patrick, R. "Biological Measure of Stream Conditions," *Sewage Ind. Wastes* 22:926–938 (1950).
20. Corliss, J. O. *The Ciliated Protozoa,* 2nd ed. (Elmsford, NY: Pergamon Press, Inc., 1979).
21. Sládecková, A. "Limnological Investigative Methods for the Periphyton ("Aufwuchs") Community," *Bot. Rev.* 28(2):286–350 (1962).
22. Paul, R. W., Jr., D. H. Kuhn, J. L. Plafkin, J. Cairns, Jr. and J. G. Croxdale. "Evaluation of Natural and Artificial Substrate Colonization by Scanning Electron Microscopy," *Trans. Am. Micros. Soc.* 96(4):506–519 (1977).
23. Yongue, W. H., Jr., and J. Cairns, Jr. "A Comparison Between Numbers and Kinds of Freshwater Protozoans Colonizing Autoclaved and Unautoclaved Polyurethane Foam Substrates," *Appl. Environ. Microbiol.* 31(5):750–753 (1976).
24. Yongue, W. H., Jr., and J. Cairns, Jr. "Long-term Exposure of Artificial Substrates to Colonization by Protozoans," *J. Elisha Mitchell Sci. Soc.* 89(1–2):115–119 (1973).
25. Cairns, J., Jr., W. H. Yongue, Jr. and N. Smith. "The Effects of Substrate Quality upon Colonization by Freshwater Protozoans," *Rev. Biol.* 10(1–4):13–20 (1974–1976).
26. Yongue, W. H., Jr., J. Cairns, Jr. and H. C. Boatin, Jr. "A Comparison of Freshwater Protozoan Communities in Geographically Proximate but Chemically Dissimilar Bodies of Water," *Arch. Protistenkd.* 115:154–161 (1973).
27. Cairns, J., Jr., J. L. Plafkin, W. H. Yongue, Jr. and R. L. Kaesler. "Colonization of Artificial Substrates by Protozoa: Replicate Samples," *Arch. Protistenkd.* 118(4):259–267 (1976).
28. Sokal, R. R., and P. H. A. Sneath. *Principles of Numerical Taxonomy* (San Francisco: W. H. Freeman & Co. Publishers, 1963).
29. Patrick, R., J. Cairns, Jr. and S. S. Roback. "An Ecosystematic Study of

the Flora and Fauna of the Savannah River," *Proc. Acad. Nat. Sci. Phil.* 118(5):109-407 (1967).
30. Simberloff, D. S. "Experimental Zoogeography of Islands: Effect of Island Size," *Ecology* 57(4):629-648 (1976).
31. Cairns, J., Jr., and W. H. Yongue, Jr. "Protozoan Colonization Rates on Artificial Substrates Suspended at Different Depths," *Trans. Am. Micros. Soc.* 93(2):206-210 (1974).
32. Patrick, R. "The Effect of Invasion Rate, Species Pool, and Size of Area on the Structure of the Diatom Community," *Proc. Acad. Nat. Sci. Phil.* 58(4):1335-1342 (1967).
33. Kuhn, D. L., and J. L. Plafkin, "The Influence of Organic Pollution Upon the Dynamics of Artificial Island Colonization by Protozoa," *Bull. Ecol. Soc. Am.* 58(2):14 (1977).
34. Cairns, J., Jr., R. L. Kaesler, D. L. Kuh, J. L. Plafkin and W. H. Yongue, Jr. "The Influence of Natural Perturbation Upon Protozoan Communities Inhabiting Artificial Substrates," *Trans. Am. Micros. Soc.* 95(4):646-653 (1976).
35. MacArthur, R. H., and E. O. Wilson. "An Equilibrium Theory of Insular Zoogeography," *Evolution* 17:373-387 (1963).
36. Cairns, J., Jr., K. L. Dickson and W. H. Yongue, Jr. "The Consequences of Nonselective Periodic Removal of Portions of Freshwater Protozoan Communities," *Trans. Am. Micros. Soc.* 90(1):71-80 (1971).
37. Cairns, J., Jr., and J. A. Ruthven. "Artificial Microhabitat Size and the Number of Colonizing Protozoan Species," *Trans. Am. Micros. Soc.* 89(1):100-109 (1970).
38. Spoon, D. M. *Survey, Ecology, and Systematics of the Upper Potomac Estuary Biota: Aufwuchs Microfauna, Phase I* (Washington, DC: Water Resources Research Center, Washington Technical Institute, 1975).
39. Cairns, J., Jr. "Protozoan Communities," in *Microbial Interactions and Communities,* A. T. Bull and A. R. Watkinson, Eds. (London: Academic Press, Inc., in press).
40. Jones, R. C., J. Cairns, Jr. and W. H. Yongue, Jr. "Vertical Gradients in Artificial Substrate-Associated Protozoan Community Structure in a Stratified Freshwater Lake," *J. Elisha Mitchell Sci. Soc.* 92(1):1-8 (1976).
41. Peet, R. K. "The Measurement of Species Diversity," *Ann. Rev. Ecol. Sys.* 5:285-307 (1974).
42. "Reviewer's Comment," Grant proposal (1977).
43. Stout, J., and J. Vandermeer. "Comparison of Species Richness for Stream-Inhabiting Insects in Trophical and Mid-Latitude Streams," *Am. Nat.* 109(967):263-280 (1975).
44. Spoon, D. M. "Use of Thin Flexible Plastic Coverslips for Microscopy, Microcompression, and Counting of Aerobic Microorganisms," *Trans. Am. Micros. Soc.* 95(3):520-523 (1976).
45. Kuhn, D. L., J. L. Plafkin and J. Cairns, Jr. "A System for Recording and Processing Taxonomic Data," *Water Resources Bull.* 14(6):1507-1510 (1978).
46. Cooke, G. D. "Experimental Aquatic Laboratory Ecosystems and Communities," in *The Structure and Function of Freshwater Microbial Communities,* J. Cairns, Jr., Ed. (Blacksburg, VA: Virginia Polytechnic Institute and State University Press, 1971), p. 47.

47. Henebry, M. S., and J. Cairns, Jr. "The Effects of Island Size, Distance, and Epicenter Maturity on Colonization in Freshwater Protozoan Communities," *Am. Midl. Nat.* 104:80–92 (1980).
48. Cairns, J., Jr., W. H. Yongue, Jr. and H. Boatin, Jr. "The Relationship Between Number of Protozoan Species and Duration of Habitat Immersion," *Rev. Biol.* 9:35–42 (1973).
49. Cairns, J., Jr., D. L. Kuhn and J. L. Plafkin. "Protozoan Colonization of Artificial Substrates," in *Methods and Measurements of Attached Microcommunities: A Review,* R. L. Weitzel, Ed. (Philadelphia: American Society for Testing and Materials, 1979), p. 34.
50. Draper, N. R., and H. Smith. *Applied Regression Analysis* (New York: John Wiley & Sons, Inc., 1966).
51. MacArthur, R. H., and E. O. Wilson. *The Theory of Island Biogeography* (Princeton, NJ: Princeton University Press, 1967).
52. Cairns, J., Jr., M. L. Dahlberg, K. L. Dickson, N. Smith and W. T. Waller. "The Relationship of Freshwater Protozoan Communities to the MacArthur-Wilson Equilibrium Model," *Am. Nat.* 103:439–454 (1969).
53. Maguire, B., Jr. "Community Structure of Protozoans and Algae with Particular Emphasis on Recently Colonized Bodies of Water," in *The Structure and Function of Freshwater Microbial Communities,* J. Cairns, Jr., Ed. (Blacksburg, VA: Virginia Polytechnic Institute and State University Press, 1971), p. 121.
54. Schoener, A. "Experimental Zoogeography: Colonization of Marine Mini-Islands," *Am. Nat.* 108:715–738 (1974).
55. Maguire, B., Jr. "The Passive Dispersal of Small Aquatic Organisms and Their Colonization of Isolated Bodies of Water," *Ecol. Monog.* 33:161–185 (1963).
56. Simberloff, D. S. "Experimental Zoogeography of Islands: A Model for Insular Colonization," *Ecology* 50:296–314 (1969).
57. Darlington, P. J. *Zoogeography: The Geographic Distribution of Animals* (New York: John Wiley & Sons, Inc., 1957).
58. Wilson, E. O. "The Nature of the Taxon Cycle in the Melanesian Ant Fauna," *Am. Nat.* 95:169–193 (1961).
59. Preston, F. W. "The Canonical Distribution of Commonness and Rarity," *Ecology* Part I 43:185–215; Part II 43:410–432 (1962).
60. Diamond, J. M. "Distributional Ecology of New Guinea Birds," *Science* 1979:759–768 (1973).
61. Yongue, W. H., Jr., and J. Cairns, Jr. "Colonization of Polyurethane Substrates by Freshwater Protozoans," *J. Elisha Mitchell Sci. Soc.* 87:71–72 (1971).
62. Yongue, W. H., Jr., and J. Cairns, Jr. "The Role of Flagellates in Pioneer Protozoan Colonization of Artificial Substrates," *Pol. Arch. Hydrobiol.* 25:787–801 (1978).
63. Henebry, M. S., and J. Cairns, Jr. "The Effect of Source Pool Maturity on the Process of Island Colonization: an Experimental Approach with Protozoan Communities," *Oikos* 35:107–114 (1980).
64. Pianka, E. R. "On r- and K-selection," *Am. Nat.* 100:463–465 (1970).
65. Gadgil, M., and O. T. Solbrig. "The Concept of r- and K-Selection: Evidence from Wild Flowers and Some Theoretical Considerations," *Am. Nat.* 106:14–31 (1972).

66. Opler, P. A., H. G. Baker and G. W. Frankie. "Recovery of Tropical Lowland Forest Ecosystems," in *Recovery and Restoration of Damaged Ecosystems,* J. Cairns, Jr., K. L. Dickson, and E. E. Herricks, Eds. (Charlottesville, VA: University Press of Virginia, 1977), p. 379.
67. Henebry, M. S., and J. Cairns, Jr. "Monitoring of Stream Pollution using Protozoan Communities on Artificial Substrates," *Trans. Am. Micros. Soc.* 99(2):151–160 (1980).
68. Plafkin, J. L. "Colonization of Artificial Substrates by Protozoa in a Small Cave Stream, Tawney's Cave, Giles County, Virginia," *Assoc. Southeast. Biol. Bull.* 23:31 (1976).
69. Kuhn, D. L., J. L. Plafkin and M. S. Henebry. Unpublished results (1980).
70. Henebry, M. S., J. Cairns, Jr., C. R. Schwintzer and W. H. Yongue, Jr. "A Comparison of Vascular Vegetation and Protozoan Communities in Some Freshwater Wetlands in Northern Lower Michigan," *Hydrobiologia* (In press).
71. Plafkin, J. L. "Colonization of Artificial Islands by Protozoa in Differing Habitats and Systems," PhD Dissertation Virginia Polytechnic Institute and State University (1979).
72. Osman, R. W. "The Influence of Seasonality and Stability on the Species Equilibrium," *Ecology* 59:383–399 (1978).
73. Cairns, J., Jr., W. H. Yongue, Jr. and R. L. Kaesler. "Qualitative Differences in Protozoan Colonization of Artificial Substrates," *Hydrobiologia* 51:233–238 (1976).
74. Osman, R. W. "The Establishment and Development of a Marine Epifaunal Community," *Ecol. Monog.* 47:37–63 (1977).
75. Taylor, W. D. "Sampling Data on the Bactivorous Ciliates of a Small Pond Compared to Neutral Models of Community Structure," *Ecology* 60:876–883 (1979).
76. Connor, E. F., and D. S. Simberloff. "The Assembly of Communities: Chance or Competition," *Ecology* 60:1132–1140 (1979).
77. Wilson, E. O. "The Species Equilibrium," in *Brookhaven Symposia in Biology,* G. M. Woodwell and H. H. Smith, Eds., No. 22 (1969).
78. Simberloff, D. S., and E. O. Wilson. "Experimental Zoogeography of Islands: a Two-Year Record of Colonization," *Ecology* 51:934–937 (1970).
79. Cairns, J., Jr. "Factors Affecting the Number of Species in Freshwater Protozoan Communities," in *Structure and Function of Freshwater Microbial Communities,* J. Cairns, Jr., Ed. (Blacksburg, VA: Virginia Polytechnic Institute and State University Press, 1971), p. 219.
80. Margalef, R. *Perspectives in Ecological Theory* (Chicago: The University of Chicago Press, 1968).
81. Odum, E. P. *Fundamentals of Ecology,* 3rd ed. (Philadelphia: W. B. Saunders Company, 1969).
82. Goodman, D. "The Theory of Diversity-Stability Relationships in Ecology," *Quart. Rev. Biol.* 50:237-266 (1975).

CHAPTER 3

ARTIFICIAL SUBSTRATES AS ECOLOGICAL ISLANDS

Richard W. Osman*

Marine Science Institute
University of California
Santa Barbara, California 93106

The construction and use of artificial substrates for experiments is really the creation of experimental islands. Because of this, it is worthwhile to consider the importance of islands in ecology and ecological studies and the role of artificial substrates as islands.

An island is simply a patch of habitat (e.g., land) separated from similar patches of habitat by a second habitat (e.g., water). As such, islands can vary in size, degree of isolation and physical structure. Their boundaries can be sharp and discontinuous, or gradational. They can vary from large oceanic islands, to islands of forest surrounded by grasslands, to trees within the forest, to leaves on the tree. What is perceived as an island will depend on the organisms being studied. However, islands are patches of habitat, defined physically and biologically. As such they are basic functional units in ecology. Almost all ecological studies consider some part or patch of the ecosystem in partial or complete isolation from all other parts. In this sense we study islands.

The organisms that inhabit an island are assumed to comprise a semi-closed system. Events that happen on the island are much more important to the inhabitants than events that happen elsewhere. Additions of species or individuals from outside can occur through immigration, and losses to the outside can result from emigration. These exchanges either

*Present address: Academy of Natural Sciences of Philadelphia, Benedict Estuarine Research Laboratory, Benedict, Maryland 20612.

can be monitored or experimentally controlled and the dynamics within the island studied. It is within islands that the restricting conditions and assumptions of theoretical models might realistically be approached. They are the logical places for the testing of ecological theories and hypotheses.

Oceanic islands that have existed for evolutionary time periods may be the sites of the only "replicated" experiments in evolution. Although no two islands can be considered to be true replicates of one another, comparisons of populations on different islands have yielded much of our knowledge concerning the mechanisms and rates of evolution.

The use of artificial substrates as islands generally has three advantages. First, an artificial substrate is similar to, but simpler than, the natural patch/island. Both the local physical and biological environment and the characteristics of the substrate can be controlled. Environmental conditions can be regulated by controlling the location and orientation of the substrate, the time of initial exposure, the length of the experiment or the species present. Substrate characteristics such as size, shape, chemical composition, physical structure and past history all can be varied or held constant. Second, with artificial substrates islands truly can be replicated. In addition, the number of possible replicates will not be constrained by the availability of natural islands. Third, the patch or island can be defined precisely by its physical boundaries. These boundaries will be sharp and known before the experiment is begun.

Since artificial substrates are excellent experimental islands, it follows that they can be used to investigate many different types of ecological hypotheses. However, there are some ecological problems in particular that are usually thought of as island questions. Probably the most well-known of these and the most studied in recent years is the species equilibrium theory of island biogeography as proposed by MacArthur and Wilson [1,2]. The models they proposed are simple extensions of birth-death equations and based on the fact that what is present on an island is the summation of positive gains and negative losses. These gains and losses are the normal dynamics of any ecological system. By making certain assumptions or hypotheses about the way in which they occur, the theory usually is used to predict the spatial and temporal pattern of species numbers that should result. In general, these are the patterns that are measured and used to test the hypotheses. Although the distributions of species among island communities are most frequently studied (e.g., by Diamond, [3,4], Lassen [5], Power [6], MacArthur and Wilson [2] and Wilcox [7]; also see Simberloff [8]), the theory can be used to predict patterns of spatial and temporal change among local populations of a single species [9,10] or changes in continental faunas over evolutionary time [11–13].

The patterns used to test the hypothesized dynamics are almost always static ones. Typically, a single survey of a particular taxon is conducted on a group of islands. This survey may be compared to a previous survey (if one exits) conducted by a different investigator, probably using different methods. Thus, few attempts have been made to measure or estimate the actual dynamic processes. Many of the assumptions of the species equilibrium theory have yet to be tested. The few strong tests of the species equilibrium theory that exist usually have involved the use of experimental islands, either defaunated natural islands [14-19] or artificial substrates.

Artificial substrates are ideal for testing many of the details of the MacArthur-Wilson model. If the appropriate taxa are chosen—ones that might be expected to colonize and equilibrate on the island within some fraction of the investigator's lifetime—and the artificial substrate can be surveyed repetitively without disturbing the colonists, then by monitoring the substrate at regular time intervals it is possible to estimate the actual dynamic processes. Immigration and extinction rates can be measured, with replication, under a variety of predetermined (e.g., island size) and/or resulting (e.g., number of species present) conditions. Also, it is possible to measure any changes in the independent variables. For example, temporal or seasonal changes in the number of species available to colonize an island can be estimated by exposing identical islands (clean substrates) for equal time periods in the different seasons. Finally, the substrates can be manipulated experimentally to test a priori predictions. For example, islands can be moved closer to a source of immigrants or increased in size.

Artificial substrates also can provide new insights into the dynamics of island communities. The processes of immigration, extinction and species interaction can be examined in great detail. Through the use of controlled experiments it is possible to examine the importance of factors that normally covary. For example, as island size increases the number of habitats on an island can be expected to increase also. Unfortunately, it is usually impossible to separate the effects of size from the effects of habitat diversity on the number of species. Simberloff [16], by dividing apparently homogeneous islands into several smaller islands, was able to examine the importance of area without habitat diversity covarying. With artificial substrates it is also possible to vary habitat diversity without changing the island size [20]. Finally, artificial substrates can also be used to examine some of the sampling problems of island ecology. Both methods and frequency of data collection often can influence the measurements obtained. The rescue effect [14] of high immigration reducing observed extinction is as much a sampling problem as it is a normal phenomenon. Schoener [21] has demonstrated that it is impossible to measure absolute rates of extinction and immigration. Relative measures also may be affected by sampling design and this effect can be demonstrated with artificial substrates.

In the following sections the results of studies of the colonization of artificial substrates by sessile marine invertebrates will be presented. The results will be used to examine the role of an artificial substrate as an ecological island, to show how artificial substrates can be used to test island biogeographic theory and, finally, to demonstrate new insights into the theory.

THE MACARTHUR-WILSON MODEL

The basic statement of the species equilibrium theory [1] is really a simple extension of a normal birth-death equation in which species replace individuals. In ecological time,

$$S_t = S_{t-1} + S_i - S_e. \tag{1}$$

where S_t is the number of species at time t, S_i is the number of new species immigrating in the time period t-1 to t, and S_e is the number of species that went extinct locally through emigration or mortality. In the particular community being examined, the organisms attach permanently to the substrate, thus eliminating emigration as a cause of local extinction.

It is obvious from Equation 1 that when S_i is equal to S_e, S_t will be equal to S_{t-1}, and the number of species on the island will be at an equilibrium. The general conditions that affect immigration and extinction rates and how these rates vary with time and place are the remaining parts of the model.

MacArthur and Wilson [1] first introduced the idea of a propagule as the smallest number of individuals of a particular taxon that colonize an island and can propagate themselves. The immigration rate then will be a function of (1) the number of propagules produced in some source region or regions; (2) the proportion of these propagules that might reach the island if all survived; (3) the survivorship of the propagules as a function of distance travelled; and (4) the proportion of the species that reach the island that are not already in residence.

MacArthur and Wilson proposed that the number of propagules produced in a source region, i, would be a function of the area of that region, A_i, and a constant, α, determined by the number of propagules produced per species by the taxa being studied. The proportion of these (αA_i) that might be expected to reach a particular island will be a function of the distance, d_i, to the island and the diameter, $diam_i$, of the island perpendicular to the direction to the source. This will be equivalent to the proportion of the circumference of a circle with radius, d_i, represented by $diam_i$ or $diam_i/2\pi d_i$. If each of these propagules has an equal chance of surviving some unit of time, then the number that would survive long enough to reach the island will be proportional to $e^{-\lambda d_i}$, where λ is a constant mortality rate. Finally, if all species are equally likely to colonize, the probability of any colonizing species being new will be equal to $1-(S/P)$, where $S=S_{t-1}$ and P is equal to the size of the species pool. Thus, for *n* source regions immigration will be

$$\sum_{i=1}^{n} \frac{\alpha A_i diam_i(1 - S/P)}{2\pi\, d_i} e^{-\lambda d_i} \tag{2}$$

From this formula it can be seen that three variables—d_i, $diam_i$ and P—will influence the immigration rate. The distance from an island to the source is usually large and is seen as the most important parameter affecting immigration. However, if an island is close to the source or the number of propagules is large, the size of an island can measurably affect immigration. Big islands or large substrates will simply sample more propagules and, on average, more species than smaller islands. The size of the species pool usually is considered to be fixed in ecological time. However, if successional processes alter the habitats of the island, then certain species may not be able to colonize until succession has occurred. These same successional changes in habitat also may exclude species that had previously been able to colonize. Thus, the size of the species pool may change with time.

Although immigration may be affected by several factors, the extinction rate should be mostly a function of population size. If p_s is the probability of mortality for an individual of species s and there are N_s individuals of species s on an island, then the probability of extinction or loss of all individuals of species s is $p_s \exp(N_s)$. The extinction rate for an island with n species will be

$$\sum_{s=1}^{n} p_s \exp(N_s) \tag{3}$$

Since population size should vary directly with island size, extinction rate should decrease as island size increases.

In addition to the above, MacArthur and Wilson [1] also predicted that immigration would decrease as species accumulated on an island and extinction would increase. The decrease in immigration arises from the simple fact that the more species present, the fewer the number of new species represented by propagules; $1-(S/P)$ will decrease as S increases. If there is a finite number of individuals that the island can hold, then the average population size of each species will decrease as species are added. The extinction rate should increase.

Because few of the parameters of the above equations actually can be measured, the equilibrium model usually is used qualitatively. Its major predictions are as follows:

1. Since the immigration rate will decrease as a function of the number of species accumulated on an island and the extinction rate will increase, the island will equilibrate at the number of species where these two rates are equal.
2. This equilibrium is dynamic. Immigrations and extinctions continue to occur and the species composition of the island will continue to change.
3. The equilibrium number of species will increase with island size. This results not only from a decrease in extinction rate with island area, but also from a possible increase in immigration rate.
4. The equilibrium number of species will decrease as the distance between an island and its source of colonists increases. This results mainly from a decreasing immigration rate with distance.

Besides these well-known predictions of the MacArthur-Wilson model, others can be made. If any phenomenon affects the immigration or extinction rates in a predictable way, its effect on the species equilibrium also can be predicted. Such predictions are not trivial. It is important to examine whether complex patterns in the distribution of species can result from simple patterns of reproduction and mortality.

THE COMMUNITY

The community of sessile animals studied includes species from more than 10 marine invertebrate phyla. The great majority of these species permanently attach to hard substrates while the remainder are sedentary, moving very little unless disturbed. There are both solitary and colonial species, but colonial forms usually dominate the substrate. The adults have a variety of external morphologies, growing as encrusting sheets, spreading vines, or upright shrubs and trees [22]. Most feed either by filtering plankton and organic particles from the surrounding water or by collecting material deposited on the surface of the substrate. A variety of motile species prey on this community, but these predators are not included in the following analyses; the association of these species with particular substrates could not be ascertained because of their motility and the nondestructive sampling methods used.

Dispersal for sessile species is usually by planktonic larvae. The larvae of different species must spend varying lengths of time in the plankton. Some larvae exist for only a few minutes and then metamorphose and attach to a substrate. Others can remain in the plankton for over a year [23]. Most of the species studied have larvae that exist for time periods of a few hours to a few weeks.

In many ways this community is very similar to a plant community. Morphologically, adults resemble miniature plants and mimic many plant growth forms. Energy is gained from an external source (through filter-feeding) and converted into local biomass. Dispersal is similar to wind dispersal of seeds, and single individuals can be propagules. The major differences are of scale. The organisms are generally smaller in size and their lifetimes often shorter. This, of course, is an experimental advantage. Several of the general characteristics of the species within this community affect the application of the island models. The most important of these are those characteristics affecting reproduction and dispersal.

For all colonial species, each metamorphosing and attaching larva is capable of asexual reproduction. For most species, the colony or solitary individual produced by a larva also will be capable of sexual reproduction, even in the absence of other individuals on the substrate. This is possible through self-fertilization or external fertilization of gametes. In this sense, a single larva can be defined as a propagule.

It is unlikely that the population size of a species on an island will increase through local sexual reproduction. Sexual reproduction, no matter where it occurs, usually results in the production of larvae. When released, most larvae actively propel themselves away from the substrate. For some species, larvae must undergo several developmental stages in the plankton before they can settle. Therefore, the product of sexual reproduction, the larva, has little probability of attaching to the substrate on which it was produced. The existence of a species on an island will be influenced more by asexual reproduction and immigration of larvae. Island size will only affect population size by its effect on immigration rate and its effect on the maximum size to which a colony can grow. The number of colonies cannot be increased by local reproduction. A distinction must be made between the populations of identical individuals or clones within a colony (the genet) and the population of genets or genetically different colonies [24]. This distinction may influence the way in which island size affects extinction rate.

METHODS

The data presented were collected as part of a three-year study of the effects of the discharged cooling water of the San Onofre Nuclear Generating Station on sessile marine invertebrates. The power plant is located on a section of open coast between Los Angeles and San Diego, California, approximately 5 km south of San Clemente, California (Figure 1). The intake and discharge for the cooling water are located approximately 800 meters offshore at a depth of about 8 meters. Permanent field stations were established at over 20 sites in the vicinity of the plant. Ten stations were placed along transects parallel to the coastline, both upcoast and downcoast from the discharge to distances of 1600 meters. Stations were also positioned in or near each of the three local kelp *(Macrocystis pyrifera)* forests; San Mateo kelp forest, 4.5 km upcoast; San Onofre kelp forest, 1-2 km downcoast; and Barn kelp forest, 11 km downcoast (Figure 1). These stations were in depths of 12-14 meters. In each kelp forest at least one station was located in the forest, one on the edge, and one about 500 meters outside the forest.

In general, experiments using artificial substrates were conducted simultaneously at all stations. Substrates consisted of square panels of two sizes, 103 cm^2 and 14.5 cm^2. These panels were attached to racks (Figure 2), which held them approximately 1.5 meters above the bottom. All panels were exposed parallel to the bottom and all measurements were made on the bottom or face-down surface. This surface was not subject to high rates of sediment accumulation, and shading greatly reduced colonization by algae.

Two types of exposure schedules were used. *Recruitment* panels were exposed for six weeks, after which they were collected and replaced by clean panels. These monitored the changes in larval recruitment at each of the stations. *Colonization* panels were exposed at each station in June, September and December 1978 and in March 1979. These panels remained in place for the length of the study. Approximately every six weeks they were brought into the laboratory in cool, aerated seawater. The species present were recorded, abundances were estimated, the panels were photographed and then all panels returned to the field. In this way the development of the community on each panel was followed for the length of the study. Two replicates were used at each station for the recruitment panels and three replicates were used for each series of colonization panels.

Identical colonization experiments thus were performed at each of the more than 20 stations. These experiments each were monitored at approximately the same times. This design allows the direct comparison of

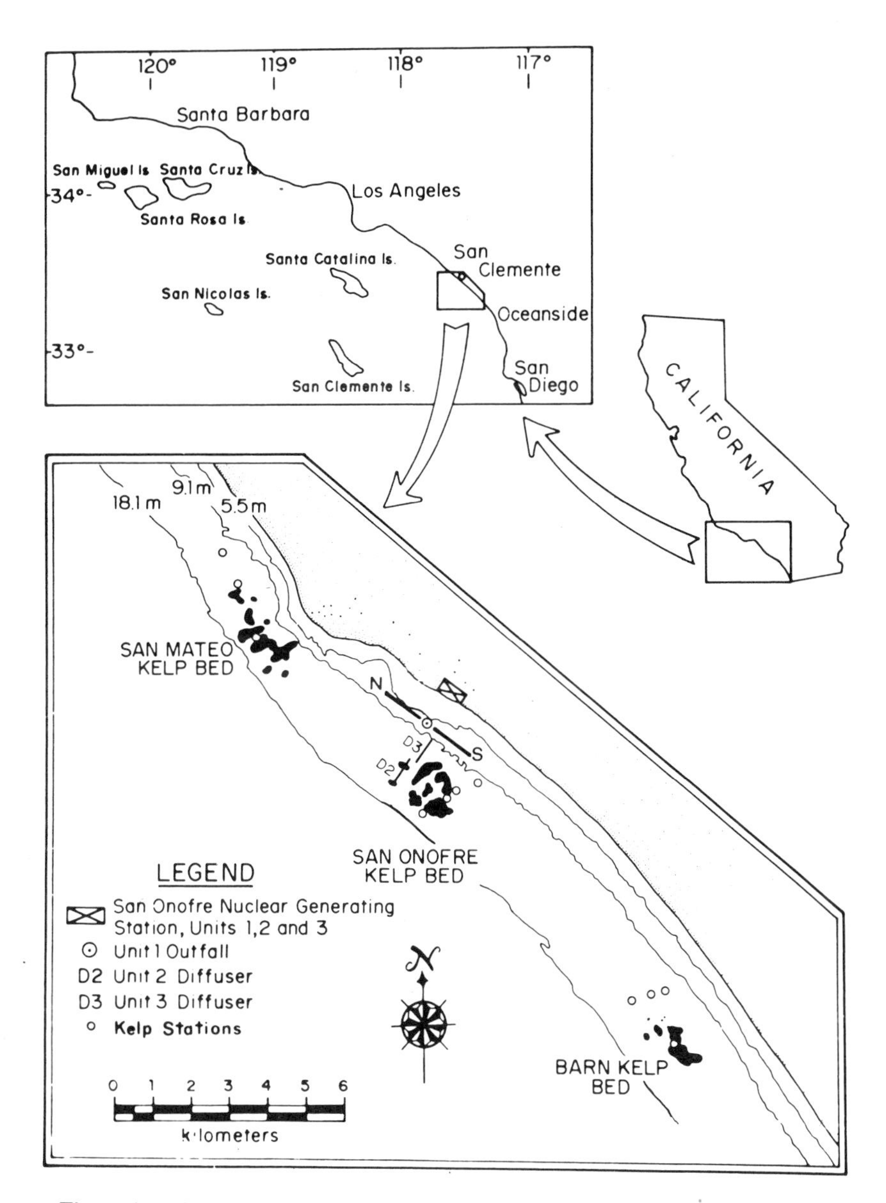

Figure 1. Map of the study area. The lines N and S indicate the locations of the north and south transects. The north transect had stations at distances of 50, 100, 200, 400, 800 and 1600 meters from the discharge (Unit 1 Outfall) of the power plant. The south transect had stations at distances of 50, 100, 800 and 1600 meters from the discharge. The black areas are the approximate locations and boundaries of the kelp forests.

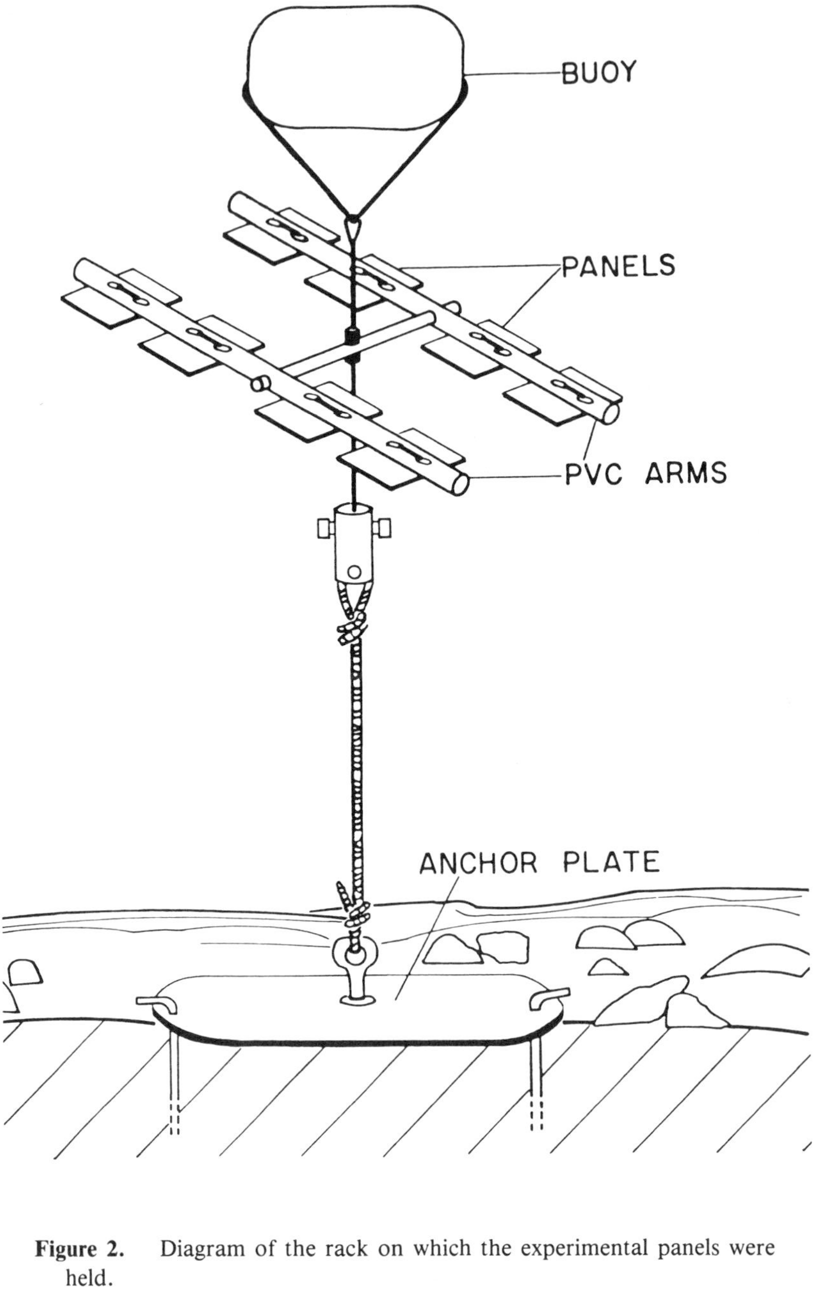

Figure 2. Diagram of the rack on which the experimental panels were held.

replicate islands, islands of different ages but in the same environment, and identical islands in different habitats.

Sampling Artifacts

The measurements used in the analyses have some degree of error introduced by the fixed sampling schedule. This error is particularly evident in the estimation of immigration and extinction rates. The estimates of these rates are based on instantaneous "snapshots" taken at widely spaced intervals; an unknown number of immigrations and extinctions are never recorded [15,21,25,26]. However, if these errors of omission can occur randomly in space and time, then the crudely measured rates still can be compared on a relative scale.

Unfortunately, the data do indicate the possibility of bias. A main source of this bias is the reciprocal relationship between extinction and immigration. The rescue effect of Brown and Kodric-Brown [14] is really only half of a more general phenomenon; if *measured* extinction is reduced by high *actual* immigration rates, then it follows that *measured* immigration will be reduced on substrates with high *actual* extinction rates. When the effects of both island size and isolation were examined, extinction and immigration were found to vary inversely (see later sections). How much the change in one rate resulted from the independent variable (size, isolation) or from the other rate (immigration or extinction) actually cannot be determined without experimentally controlling each of the variables.

These relationships between measured and actual rates also will affect the shapes of the measured immigration and extinction curves. If p_e is a known probability of a species going extinct and p_i is the probability of that same species immigrating, the measured probabilities of immigrating (I) or going extinct (E) will be

$$I = p_i - p_i p_e + p_i^2 p_e - p_i^2 p_e^2 \cdots + p_i^n p_e^{n-1} - p_i^n p_e^n$$

$$E = p_e - p_e p_i + p_e^2 p_i - p_e^2 p_i^2 \cdots + p_e^n p_i^{n-1} - p_e^n p_i^n$$

where $p_i^2 p_e$ is the probability of an immigration being followed by an extinction followed by another immigration. When p_e is 0, $I = P_i$. As p_e increases, I will decrease (Table I). Since p_e should increase as the number of species on the island increases, I will be depressed below its expected value. Obviously, p_i will have a similar effect on E, causing a depression in the measured extinction rate when only a few species are on

Table I. Probability of Observing a Species on an Island Given Known Probabilities of Immigration and Extinction[a]

	Probability of Extinction		
Probability of Immigrating	**0.90**	**0.50**	**0.10**
0.90	0.51	0.83	0.89
0.50	0.11	0.34	0.48
0.10	0.02	0.06	0.09

[a]If the known probability of immigrating is 0.90 and that of extinction is 0.50, then the probability of observing an immigration is 0.83. The table can be reversed to compute the probability of observing an extinction.

the island. It follows that the characteristic exponential relationships between measured immigration or extinction rates and the number of species present result, in part, from this relationship between the two rates.

The frequency at which substrates are sampled also will affect the measured rates [21] and their change as a function of diversity. Obviously, if the substrate is sampled at a frequency greater than the rate of arrival of propagules, the number of species immigrating during a sampling period will be 1 or 0, regardless of the number of species present. With some knowledge of the organisms being studied, it is usually possible to design an appropriate sampling schedule: daily to weekly for diatoms and protozoans [27–30]; monthly for invertebrates [21,31–36]; yearly or less frequently for birds, etc. If, however, there is a great deal of spatial variation in the rates of immigration, population growth or extinction, using the same sampling schedule at different localities may introduce bias in the measured rates. If measurements are made too frequently in some areas, it can lead to immigration rate curves with little slope.

Although sampling design can bias measurements of immigration and extinction, much of the species equilibrium model still can be tested reliably. In particular, the major prediction of an equilibrium number of species maintained by a dynamic balance between immigration and extinction should be unaffected and remain testable. This equilibrium still should change with isolation, island size and pool size. Again, hypotheses can be tested. The area in which caution must be exercised is in the interpretation of the measured immigration and extinction rates and their respective curves. It is possible for increases in extinction to be measured as decreases in immigration, and vice versa. The shape of the immigration curve may be an unreliable test of any phenomenon. Measured immigration, in particular, can be affected by many different factors, including sampling frequency. Unfortunately, there have been all too few attempts to measure immigration and extinction rates. Without data on a variety of communities or types of organisms, it is impossible to judge the importance of sampling design effects.

RESULTS AND DISCUSSION

In the following sections the results of the artificial substrate experiments will be used to address several questions concerning the distribution of species among islands. In doing this the role of the artificial substrate as an ecological island also will be examined.

Colonization Pattern and Equilibrium Numbers

If an island reaches an equilibrium in the number of species present as a result of a dynamic balance between immigration and extinction rates, it should first follow that replicate islands arrive at the same equilibrium by similar paths (in terms of species number but not species composition). Colonization curves should be similar. Each replicate is subject to the same set of potential immigrants. The size and physical environments of the islands should be identical. On average, population sizes and extinction rates should be similar. Governed by the same immigration and extinction rate curves, the replicate islands should be colonized in the same way.

In Figure 3, eight sets of replicate colonization curves are illustrated, two sets from each of four stations. These stations were chosen to represent the range of environments within the study area. Two stations are from the inshore (8 meter depth) transect, the third is from inside the San Onofre kelp forest, and the fourth is located in the San Mateo kelp forest. Except for the summer 1978 series at the San Mateo station, there is little difference between replicates within a set. In fact, there is also little difference between the summer and winter series at a station. Even though a different equilibrium may result at different stations, within a station each panel equilibrates at the same number of species. Within a set of replicates (and sometimes between sets) this equilibrium is approached at the same rate.

The panels from the San Onofre kelp forest and the two inshore sites are representative of the other stations and seasonal series. Replicate panels do have similar colonization curves. The San Mateo summer panels were the only exception. After accumulating 35–40 species, each of the panels was colonized by the colonial polychaete *Salmacina tri-*

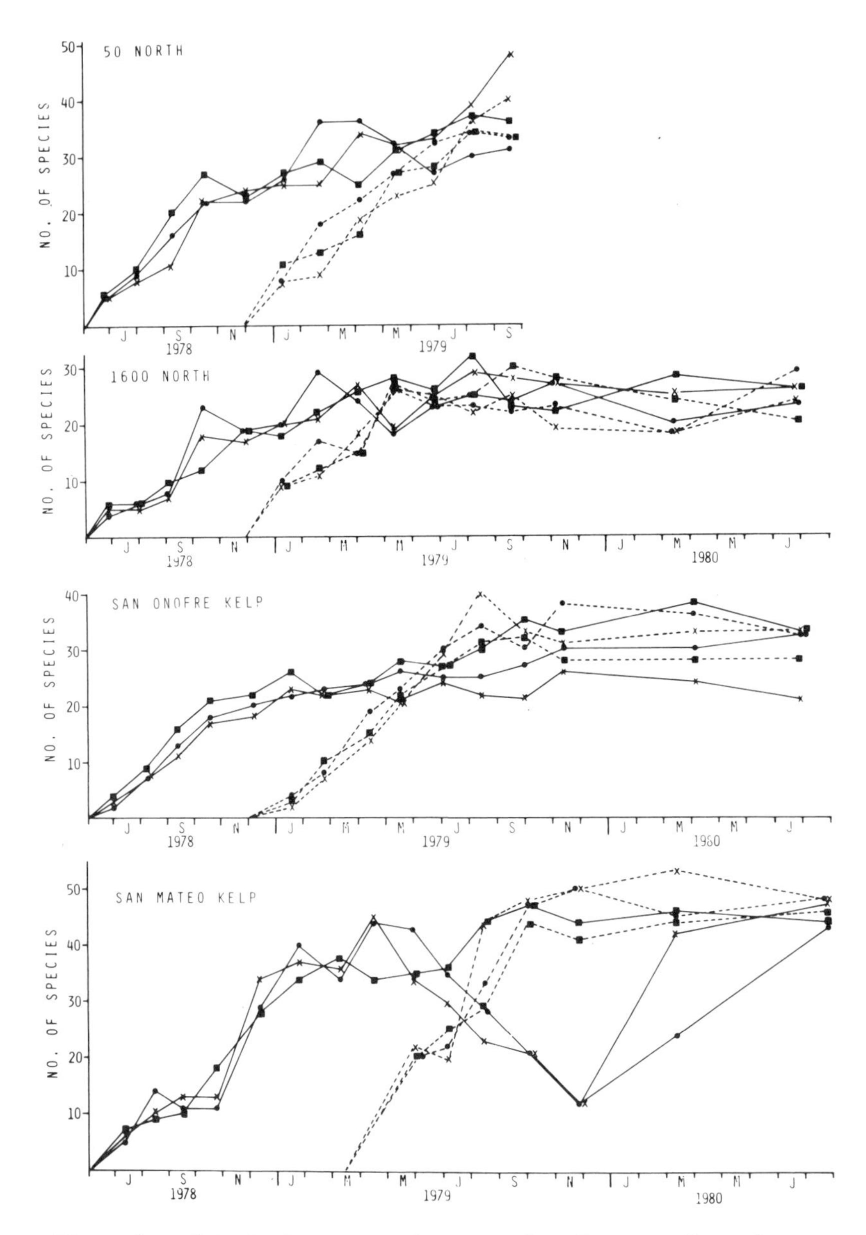

Figure 3. Colonization curves for sets of replicate panels at four stations. Each curve is for an individual panel. Solid lines are for panels initially exposed June 1978; dashed lines are for panels exposed December 1978 or March 1979 (San Mateo kelp). Sampling was discontinued at the 50 north station in October 1979.

branchiata. On two of the panels, *Salmacina* overgrew the other species and totally dominated the panels. This led to an obvious decline in the total number of species present. These fragile colonies were destroyed during winter storms, and the numbers of species increased again to over 40. By contrast, on the third panel the subdominant bryozoan *Celleporaria brunnea* was colonized by its epizoic mutual *Zanclea* sp. *Zanclea* prevented the overgrowth of *Celleporaria* by *Salmacina* [37] and, as a result, prevented the exclusive dominance of the panel by *Salmacina*. The numbers of species on this panel never fell below 40. If the spring 1979 panels at this station are representative, this is a normal equilibrium number of species.

Since immigration is a stochastic process, replicates can be expected to differ in species composition. The chance occurrence of a superior competitor on a replicate island (e.g., *Salmacina*) can increase the extinction rate, while another chance occurrence of a rare species (e.g., *Zanclea*) can reduce the extinction rate. Deterministic processes, e.g., dominance by a greatly superior competitor or the effect of a mutual, can alter the predicted equilibrium and the colonization process. Replication allows us to examine the importance of these events and their frequency.

In addition to colonization curves of replicates being similar, we also expect the colonization of islands and their equilibria to be dynamic. The species composition on an island should continue to change even though the number of species remains relatively constant. Since immigration and extinction are, at least in part, stochastic processes, as alluded to above, the species composition of replicate islands also should be different, particularly if the pool of available species is large.

Figure 4 shows the dynamic nature of two sets of panels initially exposed in autumn 1978. Relative rates of extinction and immigration were estimated as the number of species lost or gained between two sampling times ($t-1$ and t), which were plotted against the number of species present measured at the time $t-1$. The immigration and extinction curves in Figure 4 are fairly typical. The relationship between extinction and number of species present is usually strong ($r^2>0.6$), while the relationship of immigration to species present is weak ($r^2<0.3$). Since immigration is likely to be affected by other factors (Equation 3, later sections), the weak correlation to number of species present is not surprising. The colonization curves indicate that the panels did reach equilibria in the number of species present. It is obvious from the associated immigration and extinction graphs that even at equilibrium the species composition continued to change on each panel. The equilibria were dynamic.

Table II compares these same two series of replicates and a third series from the station inside the San Onofre kelp forest. When replicates from

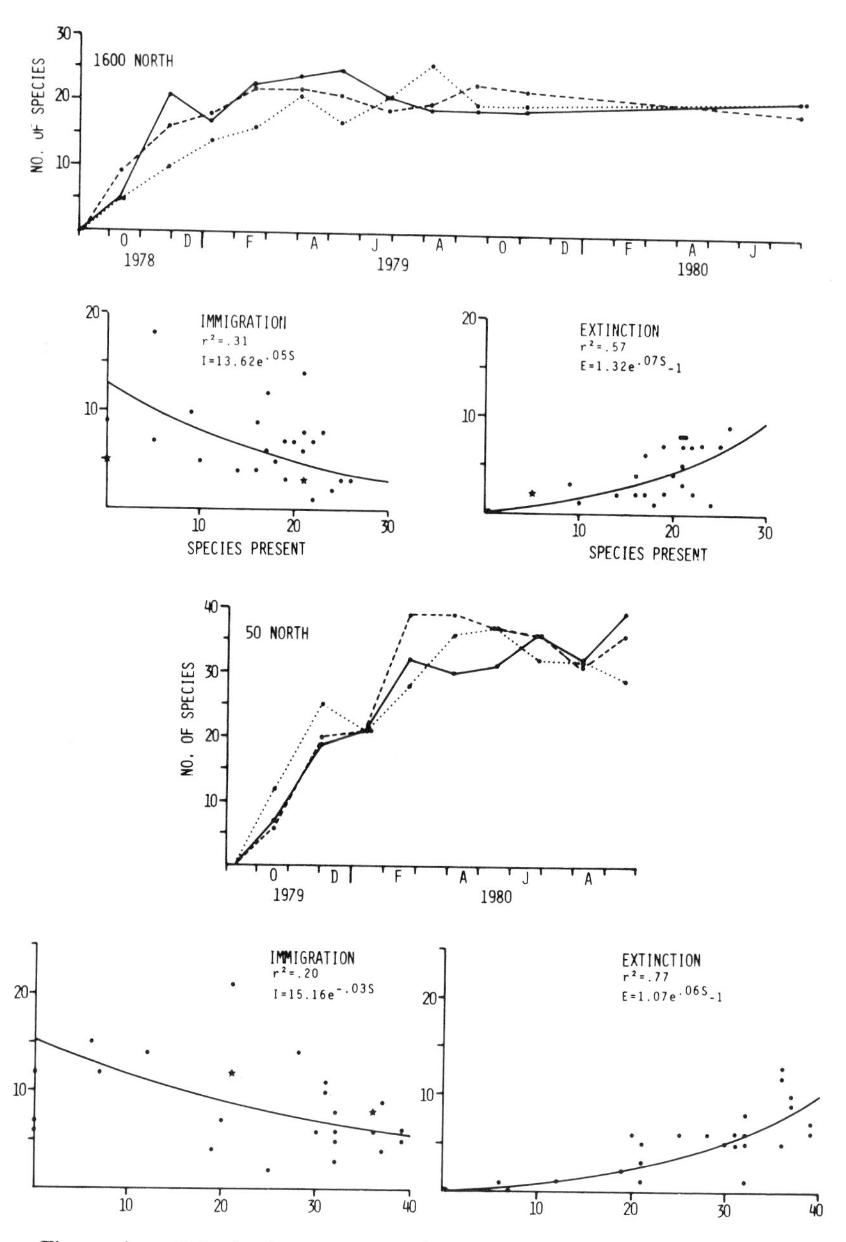

Figure 4. Colonization curves and the associated immigration and extinction curves for summer series panels at the 1600 north and 50 north stations. Immigration and extinction curves were regressed using the combined data from all three replicates. Immigration values for the initial sampling period (0 species present) were excluded from the regression because of possible bias. Stars indicate two data points.

Table II. Comparison of the Overlap in Species Composition on Replicate Panels at Three Stations[a]

	San Onofre Kelp	1600 North	50 North
Number of Sampling Periods	10	8	6
Number of Replicates per Sampling Period	3	3	3
Number of Species per Panel	25.8 ± 0.9	20.2 ± 0.5	33.0 ± 0.8
Species per Panel Found on Only 1 Replicate (%)	$22.4 \pm .01$	$23.8 \pm .03$	$16.7 \pm .01$
Species per Panel Found on Only 1 Replicate (%)	$29.4 \pm .01$	$21.2 \pm .02$	$20.2 \pm .02$
Species per Panel Found on All 3 Replicates (%)	$48.4 \pm .01$	$55.1 \pm .03$	$61.2 \pm .01$
Predicted Total Number of Species per Sampling Period—3 Replicates Combined			
1. Pool size = 150 species	65.26	52.36	78.82
2. Pool size = 250 species	68.30	54.20	83.56
Total Number of Species per Sampling Period	41.1 ± 1.6	31.9 ± 1.1	46.7 ± 1.1

[a] All panels were assumed to have reached their equilibrium number of species. Only those sampling periods after equilibrium had been reached (asymptotic part of colonization curves) were included in the analyses. San Onofre kelp panels exposed in June, 1978; the others in September, 1978.

the same sampling period are compared, 50–60% of the species are found on all replicates, 20–30% are found on two of the replicates, and approximately 20% are found on only one of the replicates. This clearly indicates that the species composition on replicate plates is not the same. Furthermore, compositional differences do not decrease with time. However, the percentage of species that are found on one, two or all three replicates does vary significantly between stations.

It should be noted that the overlap in species composition between replicate panels is much greater than what would be expected if all species were assumed to have an equal probability of being on any panel. The expected total number of species for all three replicate panels sampled at one time can be calculated using the equation $3S-3(S^2/P)+(S^3/P^2)$ [17], where S is the equilibrium number of species and P is the size of the species pool. Using the mean number of species per panel at equilibrium at each station as S and estimating the species pool as 150–250 species (the range of estimates based on both the total number of species found per station and the total number of species found throughout the study), an expected number of species to be found on all three replicate panels was calculated. As can be seen in Table II, the observed totals are 50–60% of the expected. The expected values are calculated assuming each species has an equal probability of being present on an island. This assumption is usually violated, creating a greater overlap in species composition on the replicate islands. For example, equal numbers of propagules are not produced by all species in a source region. Species will vary in their probabilities of immigrating and, thus, of being present.

Island Size

The effect of island size on the number of species present has been the subject of many studies. As Gilbert [38] has stated: "The strictly delimited boundaries of oceanic islands, and the comparative ease of making counts of the total number of species, have made the method of demonstrating a species-area relationship the most popular means of providing evidence for the theory, and the number of such studies is great." However, the demonstration of such a relationship is not sufficient as proof of the model [38,39].

Several predictions of the species equilibrium theory regarding island size can be made and tested:

1. Immigration rate will increase with increasing island size.
2. Extinction rate will decrease with increasing island size.
3. As a consequence of the above, large islands should equilibrate at a greater number of species.

Several studies [30,33–35,40,41] have demonstrated that the numbers of species that accumulate on artificial substrates vary directly with the size of the substrate. In a previous study [33,34] it was demonstrated that the numbers of species on substrates equilibrated at different values, depending on substrate size. The greater number of species on large panels was largely a result of increased immigration rates. The predicted difference in extinction rate was not observed.

In the present study, large panels consistently reached higher equilibrium numbers of species than small panels. Figure 5 compares colonization curves for identical series of large and small panels at four different stations. It can be seen that the colonization curves at different stations are asymptotic at different numbers of species. However, regardless of these station-to-station differences, small panels always accumulate a lower number of species than their companion large panels.

The effect of island size on immigration and extinction rates also was examined. Immigration and extinction curves (as in Figure 4) for companion large and small panels were compared using analysis of covariance [42]. The results of these analyses for summer series at five stations are shown in Table III. Unlike the results of Osman [34], there were significant differences between the extinction rate curves for small and large panels. As expected, the slopes of the extinction curves were much greater for small panels than for large ones; with an equal number of species present, extinction will be higher on small panels than on large.

As predicted by Equation 3, the size of an island also has an effect on the immigration rate. In each of the series presented in Table III, the number of species immigrating (i.e., the y-intercept of the immigration rate curve) is significantly greater on large panels. These results agree with the findings of Osman [34]. However, unlike the results of the earlier study [34], no significant effect was found of the number of species present on the immigration rate (slopes were not significantly different from zero). This is a major contradiction of the equilibrium theory; immigration should decrease as the number of species increases. The absence of a significant regression, particularly for the summer 1978 series, results from lower than expected immigration rates during the first three sampling periods. The events causing this departure from the expected relationship are discussed later.

Isolation

Increased distance from sources of immigrants is one of several causes of isolation, and the MacArthur-Wilson model uses distance as a measure of this isolation (Equation 3). The model predicts that the number of

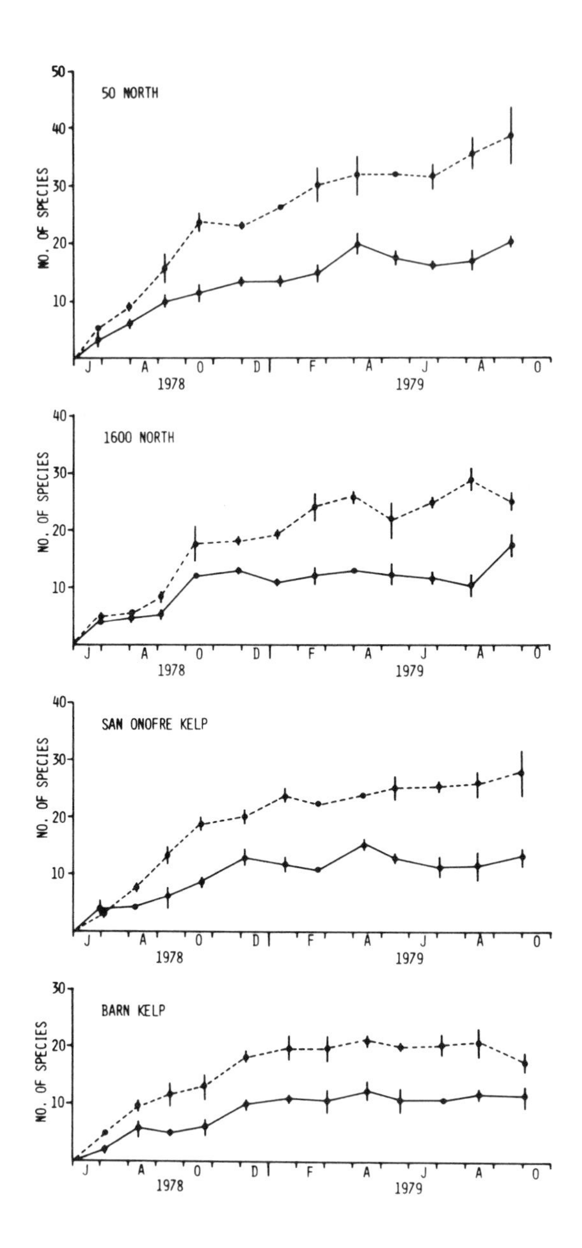

Figure 5. Comparison of colonization curves for large and small panels exposed at four different stations. All panels were exposed initially in June 1978. Dashed lines are for large panels and solid lines are for small panels. Each point is the mean of three replicate panels and the bar represents the standard error of the mean.

propagules reaching an island and, on average, the number of species immigrating, will decrease with increasing distance. With lower immigration rates, distant (isolated) islands should equilibrate at fewer species. In addition, Brown and Kodric-Brown [14] have argued that some species will be maintained on islands close to a source by continual immigration. They suggest, therefore, that the measured extinction rate will increase with the distance from a source. This should magnify the expected decrease in equilibrium number with increasing distance from a source.

The effects of isolation on immigration, extinction and the equilibrium diversity all can be examined using artificial substrates. If a fairly exclusive source area can be identified, substrates can be placed at increasing distances from it and the effects measured. In addition, with artificial substrates it is also possible to transplant islands from one locale to another. If the expected equilibrium number at each location is known, the increase or decrease to a new equilibrium number should be predictable. Such predictions can be tested.

The biology of the organisms studied will affect the degree of isolation represented by particular distances. The majority of the sessile marine invertebrates studied have planktonic larval stages that exist for time periods of minutes to over a year. Maximum dispersal should be some function of the lengths of these states. In addition, the larvae of many species must remain planktonic for some minimum time period to complete their development. Also, most species of larvae swim toward the water surface on being released and, in time, move back toward the bottom. This behavior is often a response to light and/or gravity [43] and tends to ensure wider dispersal. However, both the necessary development time and the larval behavior mean that a larva is often capable of immigrating onto a substrate only after some finite period of time.

Thus, for many species, the distances between an island and identified source regions may reflect inaccurately the potential immigration from that source. On the other hand, for species with short-lived larvae (minutes to hours), distance from an obvious source still may have an effect on immigration rate, particularly if current velocities are low.

I have attempted to test the hypothesized effects of isolation using several different sets of data. Two obvious source areas of immigrants were identified, and substrates placed at various distances from these sources. Kelp forests (Figure 1) are one type of source area. In the offshore zone (12–14 meters depth) most rock and cobble reefs are associated with kelp forests. The remaining bottom usually is covered by sediment. Obviously the reef areas are the major sources of potential immigrants for invertebrates that live in rocky habitats. Stations were established inside kelp forests, on the edges of forests and at distances of 0.5 km from them.

Table III. Summary of Analyses of Covariance Comparing Immigration or Extinction Rates for Large and Small Panels Exposed June 1978[a]

Station	r^2	a	b	Comparison of Slopes		Comparison of Y-Intercepts	
				F	p	F	p
Extinction Rate				df = 1,68			
San Onofre Kelp-In							
Large panels	0.66	1.58	0.057	8.60	<0.005		
Small panels	0.62	1.23	0.097				
San Mateo Kelp-In							
Large panels	0.60	1.23	0.047	11.25	<0.005		
Small panels	0.47	1.12	0.099				
Barn Kelp-Out							
Large panels	0.58	1.60	0.062	7.68	<0.01		
Small panels	0.59	1.10	0.115				
1600 North							
Large panels	0.75	1.05	0.074	5.08	<0.05		
Small panels	0.67	1.16	0.115				
50 North							
Large panels	0.68	1.23	0.055	11.25	<0.005		
Small panels	0.75	1.12	0.099				
Immigration Rate				df = 1,62		df = 1,62	
San Onofre Kelp-In							
Large panels	0.08	6.32	0.005	1.57	>0.1	16.63	<0.001
Small panels	0.04	4.90	−0.019				

San Mateo Kelp-In							
Large panels	0.0001	7.19	−0.00004	0.01	>0.75	7.22	<0.01
Small panels	0.0007	4.84	−0.0015				
Barn Kelp-Out							
Large panels	0.12	7.41	−0.018	2.02	>0.25	15.29	<0.001
Small panels	0.01	3.79	0.0009				
1600 North							
Large panels	0.02	4.86	0.009	0.44	>0.50	5.51	<0.01
Small panels	0.004	4.01	−0.01				
50 North							
Large panels	0.03	7.00	0.006	0.002	>0.75	13.14	<0.001
Small panels	0.003	4.52	0.005				

[1]The general equation used was an exponential with X = species present + 1 and Y = extinction or X = species present and Y = immigration. For the immigration analyses X-values of 0 (first sampling period) were excluded. Comparisons of Y-intercepts were not made if the slopes were significantly different.

A second type of source area was the area near the discharge of the power plant. Approximately 1.33×10^6 liter/min of water are discharged through a vertical pipe when the plant is operating. In addition, approximately five times this volume is drawn from the surrounding water and pulled toward the surface creating a mini-upwelling area [44]. The process results in increased larval recruitment to artificial substrates placed in this area. This increase probably results from some combination of (1) the increased local water flow causing a substrate to be exposed to a greater volume of water and, thus, more larvae; (2) increased densities of larvae in a local recirculating cell, resulting from upwelling and subsequent movement of larvae towards the bottom; and (3) stimulation of larval metamorphosis by local turbulence and/or chemicals contained in the discharge water. Whatever the mechanism, the area near the discharge is functionally less isolated in that substrate islands are exposed to a greater number of propagules and immigrating species.

As predicted, the number of propagules reaching each artificial substrate decreases with distance from a kelp forest. Table IV shows the mean number of individuals (all species combined) found on recruitment panels (immigration with zero species present). The means were significantly higher at stations inside kelp forests than those 0.5 km away. Stations at the edges of kelp forests were intermediate. The trend for number of species is similar, but the differences are not as clear. Only one outside station is significantly different from the two kelp forest stations. Differences in the equilibrium number of species are also as expected. Figure 6 shows the colonization curves for panels at each of the San Mateo kelp forest stations. Although the panels from inside the kelp forest fluctuate, it is clear that the number of species per panel decreases from inside to outside of the forest. Finally, there is some evidence for an increase in the measured extinction rate at the stations removed from the kelp forests. Between panels of the same size, the slopes of the extinction curves are significantly higher at outside stations. An example is shown in Table III. The slopes for both large and small panels are lower at the stations inside the San Onofre and San Mateo kelp forests than outside the Barn kelp forest.

If one assumes that the reef areas associated with kelp forests are source areas, the proximity of substrates to these reefs does seem to have a measurable effect on immigration rates and on the equilibrium number of species. It is possible that extinction also may be affected.

The results for the transects away from the power plant discharge are very similar to those for the kelp forests. Table V compares the recruitment of propagules and species at stations along the north and south transects. In general, as the distance from the discharge increases, the

Table IV. Analysis of Recruitment at Kelp Forest Stations[a]

Station	Position in Kelp Forest	Mean (±SE)	Stations Showing No Significant Difference
Individuals			
San Mateo Kelp 3	Outside	12.36±5.82	
Barn Kelp 2	Outside	33.25±14.21	
Barn Kelp 3	Outside	29.60±11.21	
San Onofre Kelp 3	Outside	27.50±8.93	
San Onofre Kelp 4	Edge	29.74±6.76	
Barn Kelp 1	Outside	26.42±6.56	
San Mateo Kelp 2	Edge	50.29±12.89	
San Onofre Kelp 2	Edge	31.46±6.90	
San Onofre Kelp 1	Inside	77.67±15.12	
San Mateo Kelp 1	Inside	130.13±47.28	
Species			
San Mateo Kelp 3	Outside	3.55±0.67	
Barn Kelp 2	Outside	4.15±0.54	
Barn Kelp 3	Outside	4.30±0.74	
San Onofre Kelp 4	Edge	4.39±0.57	
San Mateo Kelp 2	Edge	4.71±0.49	
Barn Kelp 1	Outside	5.17±0.56	
San Onofre Kelp 3	Outside	5.18±0.56	
San Onofre Kelp 2	Edge	5.50±0.61	
San Mateo Kelp 1	Inside	5.52±0.67	
San Onofre Kelp 1	Inside	5.71±0.38	

[a]Recruitment is expressed as total number of individuals or total number of species per panel. Vertical lines connect stations showing no significant differences based on Duncan's new multiple-range test and a two-way ANOVA (station by sampling time). Total individuals were log-transformed for analyses and resulted in some changes in the ordering of stations relative to the arithmetic mean.

mean number of propagules found on clean substrates decreases. As in Table IV, the trends for species immigration are similar but not as strong. It should be noted that there are some anomalies along the north transect. Particularly, the recruitment at the station 800 meters north of the discharge is much higher than expected. There are scattered cobble patches to the north of the discharge and this station is located in the center of one. The effects of these cobble source areas probably confound the results. To the south of the discharge there are no cobble areas, and the pattern of decreasing recruitment with increasing distance from the discharge source is much clearer. The general decreases in number of propagules and immigrating species with increasing distance from the discharge mimic changes we associate with increasing isolation. As expected, the equilibrium number of species on panels decreases with

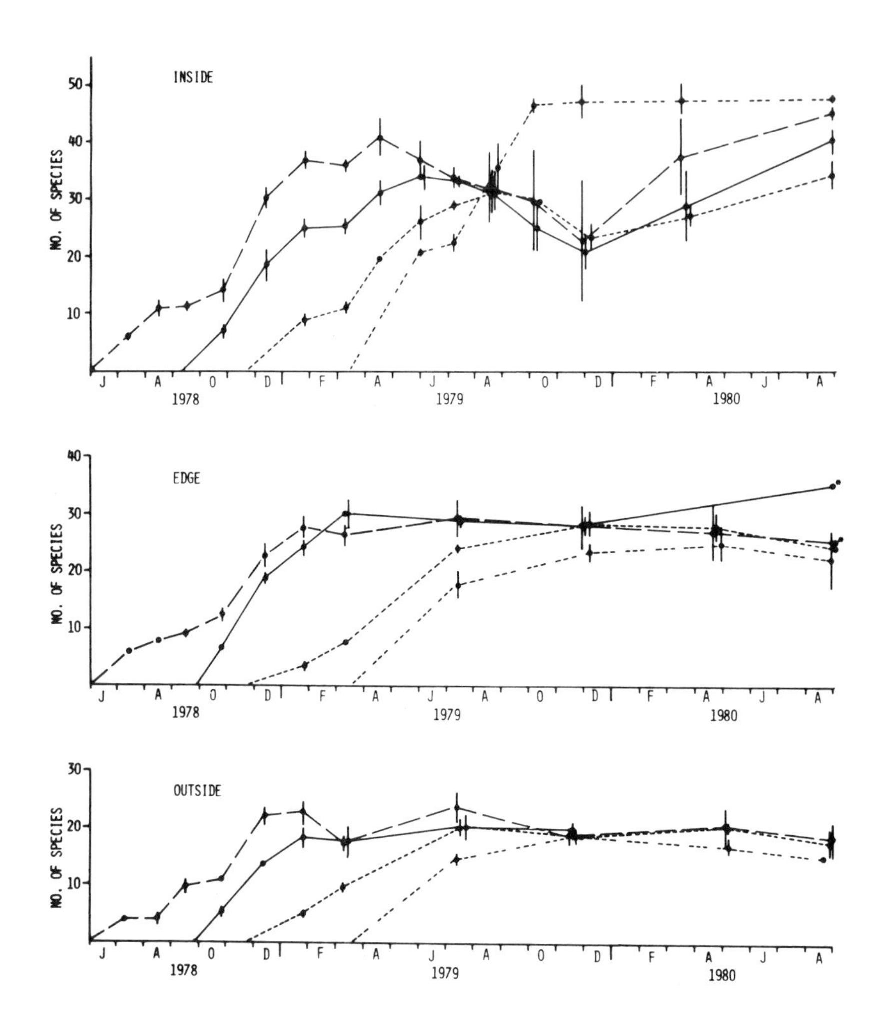

Figure 6. Comparison of colonization curves for large panels exposed at the three San Mateo kelp forest stations. The stations were inside the forest, on the edge and approximately 500 meters outside the forest (see Figure 1). All four series of colonization panels are shown for each station; panels were exposed initially in June 1978 — — — —, September 1978 ———, December 1978 ----- and March 1979 --- ---. Each point is the mean of three replicate panels, and the bar represents the standard error of the mean.

Table V. Analysis of Recruitment Along Inshore Transects[a]

Station	Location Along Transect	Mean (±SE)	Stations Showing No Significant Difference
Individuals			
1600 South	Far	20.08±3.25	
1600 North	Far	43.29±10.73	
800 South	Far	46.00±8.23	
400 North	Intermediate	76.32±16.15	
200 North	Intermediate	80.67±20.60	
50 South	Near	86.71±17.04	
50 North	Near	95.54±15.51	
800 North	Far	138.82±36.13	
100 South	Near	98.46±17.07	
100 North	Near	165.32±42.14	
Species			
1600 South	Far	5.21±0.65	
800 South	Far	6.94±0.80	
1600 North	Far	7.12±0.78	
200 North	Intermediate	7.29±0.55	
50 South	Near	7.29±0.71	
400 North	Intermediate	8.59±1.01	
100 South	Near	8.91±1.19	
50 North	Near	9.87±1.06	
800 North	Far	10.32±1.57	
100 North	Near	10.59±1.21	

[a] Numbers of stations and compass direction correspond to distance in meters and approximate direction from the power plant discharge (see Figure 1). Analyses were the same as in Table IV.

distance from the discharge. This is shown in Figure 7. In addition, colonization curves for various seasonal series of substrates at the 50-meter and 1600-meter North and South stations can be compared in Figures 3, 4, 5 and 8. Colonization curves at the 50-meter stations consistently show a higher equilibrium number of species than those at the distant 1600-meter stations. As in the kelp forest example, extinction rates seem to increase slightly with distance. This is suggested by Table III and Figure 4.

Although the area near the discharge of the power plant is not a true source area, it functions as one. Whether immigration is increased by higher densities of propagules or by some stimulation of metamorphosis causing an increased probability of settlement, the result is the same. The discharge area may not be the source area from which propagules originally were released, but rather it is an identifiable source area of competent, ready-to-attach larvae. Just as with true isolation, the number of

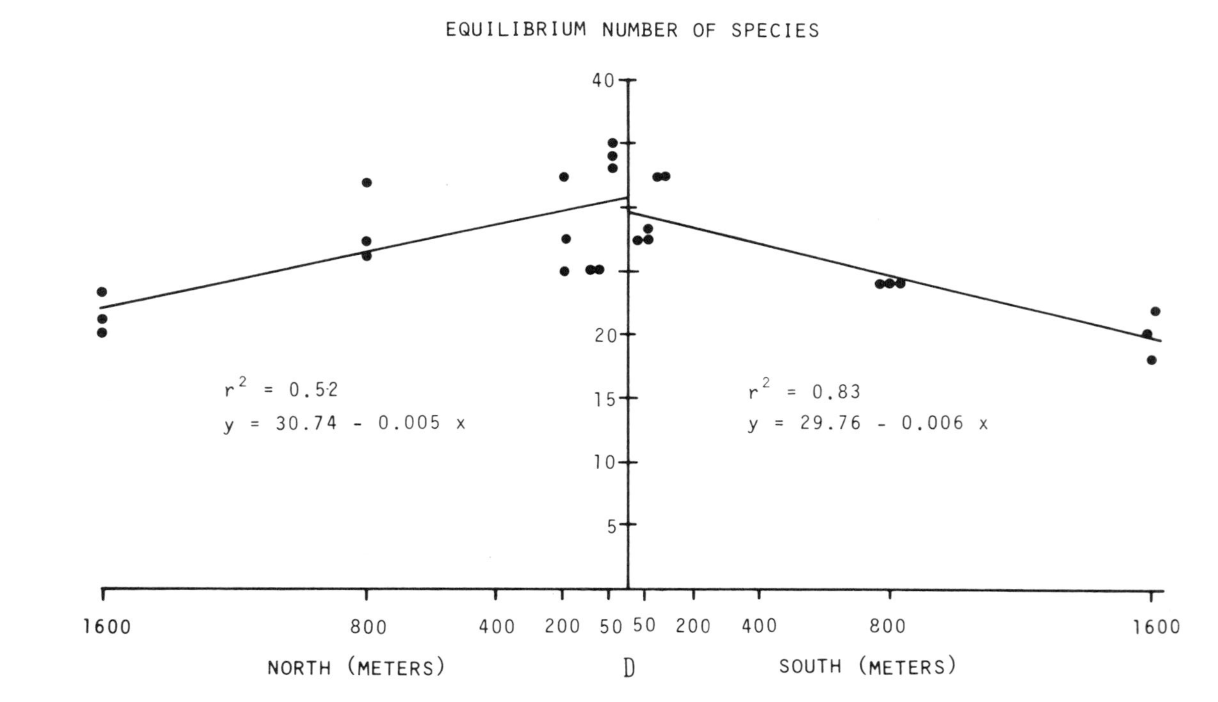

Figure 7. The relationship between the equilibrium number of species on colonization panels and the distance the panels were from the discharge of the power plant. The equilibrium number was estimated for each seasonal series of panels. Only those series that had reached equilibria were used. In the regressions, the north and south transects were treated separately, with distance always being positive.

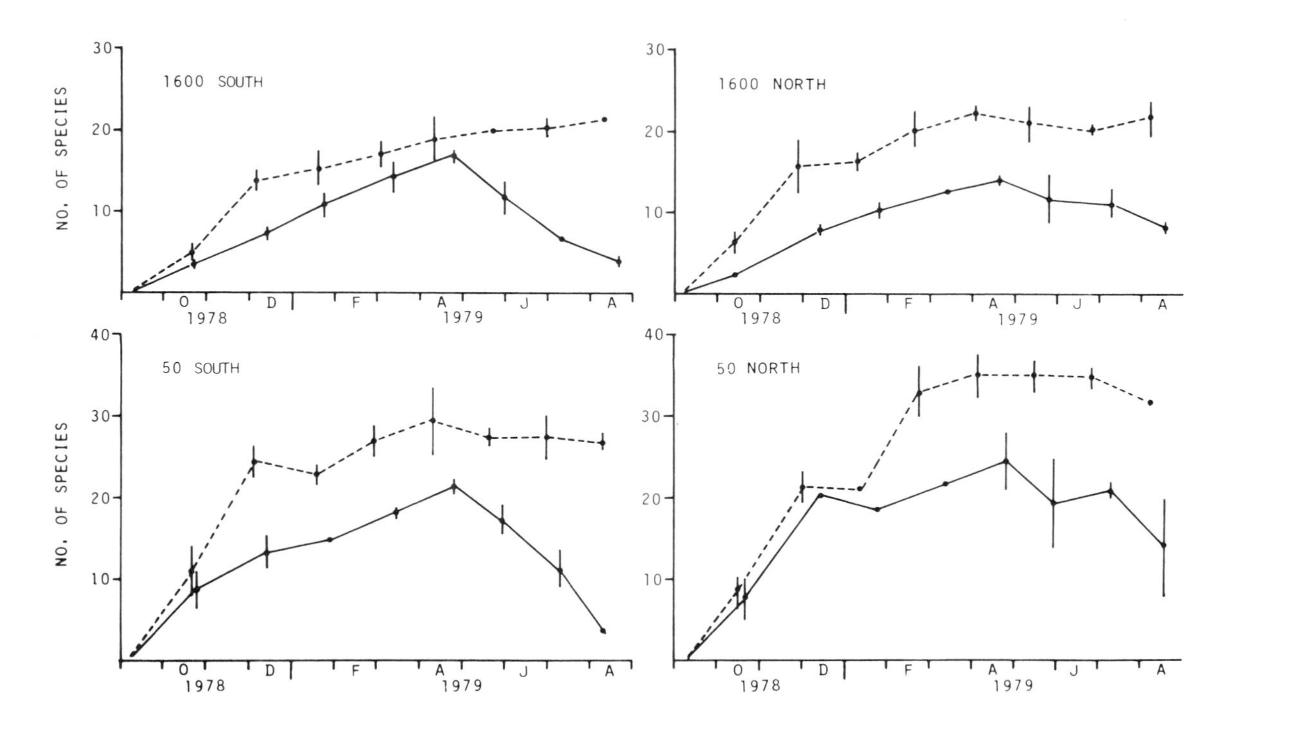

Figure 8. Comparison of colonization curves for large panels exposed at surface and bottom locations at four different stations. All panels were exposed initially in September 1978. Colonization curves for surface panels are represented by a solid line and bottom panels by a dashed line. Each point is the mean of three replicate panels, and the bar represents the standard error of the mean.

propagules a substrate island is exposed to decreases with distance from this source. The effect of this decrease on the equilibrium number of species is as predicted.

The behavior of larvae at the time of metamorphosis also can isolate substrate islands in particular locations. This type of isolation can be just as effective as distance from a source. Larvae that are ready to attach, move from surface waters toward the seafloor. Thus, substrates placed near the water surface should encounter fewer propagules, have lower immigration rates and equilibrate at fewer species. At four inshore stations panels were hung from surface buoys about 1 meter below the water surface. Table VI compares recruitment panels from these surface locations to their companion bottom panels. At each station the number of propagules (individuals) and the number of species immigrating (with zero species present) were higher for bottom panels. Differences between bottom and surface panels were significant at three of the four stations (1600N, 50N and 50S for total number of individuals and 1600N, 1600S and 50N for number of species). In addition, the eight locations follow a sequence that might be expected if both their surface-bottom locations and distance from the discharge source are considered. The panels with the lowest recruitment are at surface locations far from the power plant discharge and the ones with the highest recruitment are bottom panels near the discharge. The panels from the other locations are intermediate.

The effect of reduced immigration near the surface on the equilibrium number of species can be seen in Figure 8. At all four stations, surface panels consistently had fewer species than bottom panels. However, the surface panels did not equilibrate before the number of species began to decline precipitously. At the end of the experiment, this decline in species number occurred on all surface panels. This decline resulted from dominance by the mussel, *Mytilus edulis.* In fact, it was the mass of these mussels that ended the experiment by sinking the buoys and attached panels.

A final test of the effect of isolation on the equilibrium diversity can be made by transplanting panels from one location to another. If the two areas have a known difference in immigration or extinction rates, the resultant change in species diversity on the transferred panels should be predictable. During the course of this study, several sets of panels were transplanted from one station to another. Unfortunately, these transfers were not made to test the equilibrium model; transplants were in one direction only, from stations with low immigration rates to stations with high immigration rates. Figure 9 shows the results of transplanting substrates taken from a station outside of the Barn kelp forest to a location inside the forest. The colonization curves for two sets of these trans-

Table VI. Analysis of Recruitment at Surface and Bottom Stations Close to and Far from the Power Plant Discharge

Station	Location	Mean	Stations Showing No Significant Difference
Individuals			
1600 North	Surface	5.1	
1600 South	Surface	18.6	
50 South	Surface	19.8	
1600 South	Bottom	20.1	
50 North	Surface	38.7	
1600 North	Bottom	43.3	
50 South	Bottom	86.7	
50 North	Bottom	95.5	
Species			
1600 North	Surface	2.7	
1600 South	Surface	3.3	
50 North	Surface	5.1	
1600 South	Bottom	5.2	
50 South	Surface	6.2	
1600 North	Bottom	7.1	
50 South	Bottom	7.3	
50 North	Bottom	9.9	

[a] Numbers of stations and compass direction correspond to distance in meters and approximate direction from the discharge (see Figure 1). Analyses were the same as in Table IV.

planted substrates are compared to curves for panels of the same age left in place outside of the forest. It can be seen that up to the time of transfer there is little difference between the two colonization curves. However, immediately after transplanting, the panels placed inside of the kelp forest increased significantly in species number, while the control panels that were kept outside of the forest remained at the same equilibrium number of species as before the transplants. This immediate increase in species number when substrate islands are moved closer to a source of immigrants is exactly what is predicted by the model.

The above analyses all indicate that the effects of isolation, particularly as measured in distance from a source of immigrants, are as hypothesized in the MacArthur-Wilson model. The number of propagules that reach an island do indeed decrease with increasing isolation from a source. This usually results in lower species immigration rates and lower equilibrium numbers of species. There is also some indication that measured extinction rates may increase with increasing isolation. The results presented herein are consistent, even though the manner of isolation was different in each example. Finally, the experimental transplanting of

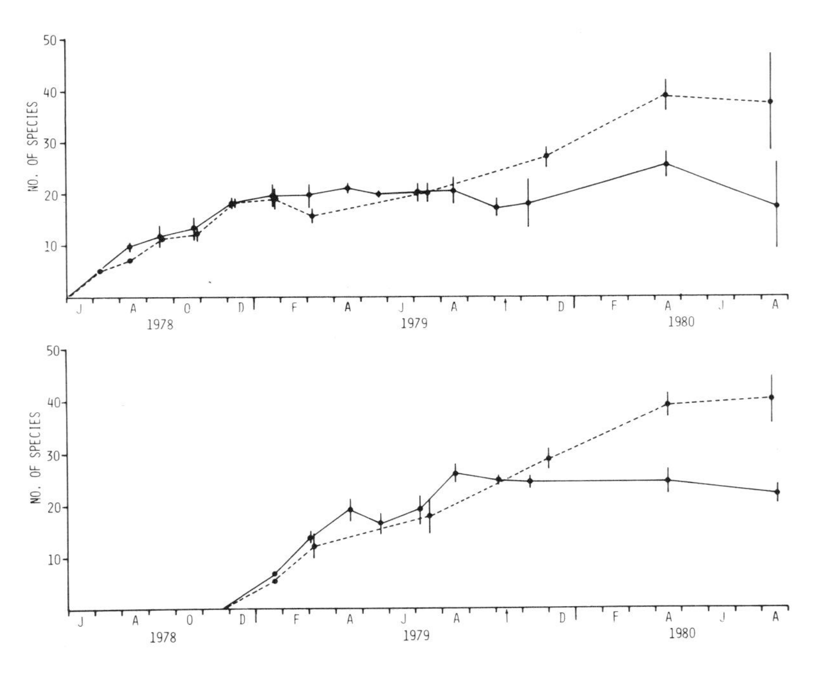

Figure 9. Comparison of colonization curves for transplanted and control panels. All panels were exposed initially at one of the three stations outside of the Barn kelp forest (see Figure 1). In October 1979 (time indicated by the arrow), panels from the station nearest the shore were transplanted into the Barn kelp forest. Control panels at most offshore station were not moved. Dashed lines are the transplanted panels and solid lines are the controls. The top graph shows panels initially exposed in June 1978 and the bottom graph shows panels initially exposed in December 1978. Each point represents the mean of three replicate panels and the bar is the standard error of the mean.

substrate islands at equilibrium in one location to another location where the equilibrium value was higher demonstrates that these equilibria are dynamic. If either the immigration or extinction rates change, the diversity of the island will change accordingly.

Seasonal Changes in Immigration

The MacArthur-Wilson model assumes no temporal fluctuation in the number of immigrants. However, it is more realistic to expect the immigration rates for most groups of organisms to vary in time; the production of propagules will seldom be constant. For many marine invertebrates peaks in larval abundance can occur at both regular and irregular intervals. If a sufficient number of species have reproductive cycles that covary, then there may be a measurable seasonal effect on the number of species that immigrate. The results of such temporal changes are no different from those of the transplant experiments. If the immigration rate increases, the equilibrium number of species also should increase, whereas decrease in immigration should cause a reduction in the equilibrium.

In a previous colonization study [33,34], a highly seasonal climate resulted in maximum immigration rates in the late summer with decreases to nearly zero in the winter. There were two major consequences of this fluctuating immigration rate. First, the shape of the initial part of a colonization curve varied greatly, depending on the time of year that a substrate was initially exposed [34]. Secondly, the expected equilibrium values varied from a maximum in the late summer to a low in the late winter. Accordingly, when a substrate reached an "equilibrium," the number of species present increased and decreased seasonally.

This continual fluctuation in diversity was not observed in the present study. There were significant changes in larval recruitment through time, with many species reproducing and releasing larvae only during particular times of the year. However, few species apparently covaried in their reproductive seasons so that a general seasonal cycle was not evident. As a result, most colonization curves appeared to be asymptotic (Figures 3-6, 8).

Immigration and extinction rates can vary both in space and time. Most studies have examined spatial variation, particularly the distance from a source of immigrants. For most islands, location and size are static and the effects of these variables relatively easy to predict. Temporal changes in immigration and extinction are equally important, but less predictable. For example, knowing that species reproduce seasonally is not sufficient for predicting a seasonal variation in species number.

Rather, enough species must covary to affect measurably the immigration rate, and changes must occur over a sufficient length of time to affect *measurably* the species equilibrium. Because of the usual absence of such data, temporal variation never has been included in the equilibrium model.

Changes in the Species Pool

The size of the pool of species available that can colonize an island will affect the immigration rate (Equation 3) and, in turn, the equilibrium number of species. As the pool size increases, $1-(S/P)$ also will increase. With a larger species pool, a greater proportion of the species immigrating onto an island with S species already present will be new species. Thus, the immigration rate can be expected to decrease more slowly when the species pool is larger.

In previous studies, the pool size has been assumed to be constant in ecological time. However, for many groups of organisms the pool size might vary as a function of the "state" of the island. During the colonization of an island, successional sequences may result in changing pool sizes. If a group of species cannot possibly immigrate onto an island until a particular habitat is present, then those species initially cannot be considered part of the pool. If other species eventually change the island or create a special habitat, then these excluded species can immigrate and the species pool will have increased. (A loss of species from the pool also could result from successional changes. No evidence has been found for this occurring on experimental panels.) If such successional changes in the species pool occur gradually during the colonization of an island, they may be difficult or impossible to measure. On the other hand, if an event causes a large increase in pool size (e.g., a doubling), then the shapes of the immigration and colonization curves may be altered measurably. This is illustrated in Figure 10. An increase in pool size can counter the normal decrease in immigration rate. As species accumulate, immigration may remain constant or even increase initially [45]. With a peak in immigration at some intermediate diversity, the rate of colonization would increase when diversity reaches this level. A constant rate of immigration would result in a slower approach to equilibrium.

Changes in pool size for artificial substrate islands may occur in various ways. First, many sessile invertebrates require the presence of a bacterial/diatom film on the substrate before they will attach. This almost always leads to reduced numbers of species immigrating during the first sampling period after the initial exposure of a panel (Figures 4,

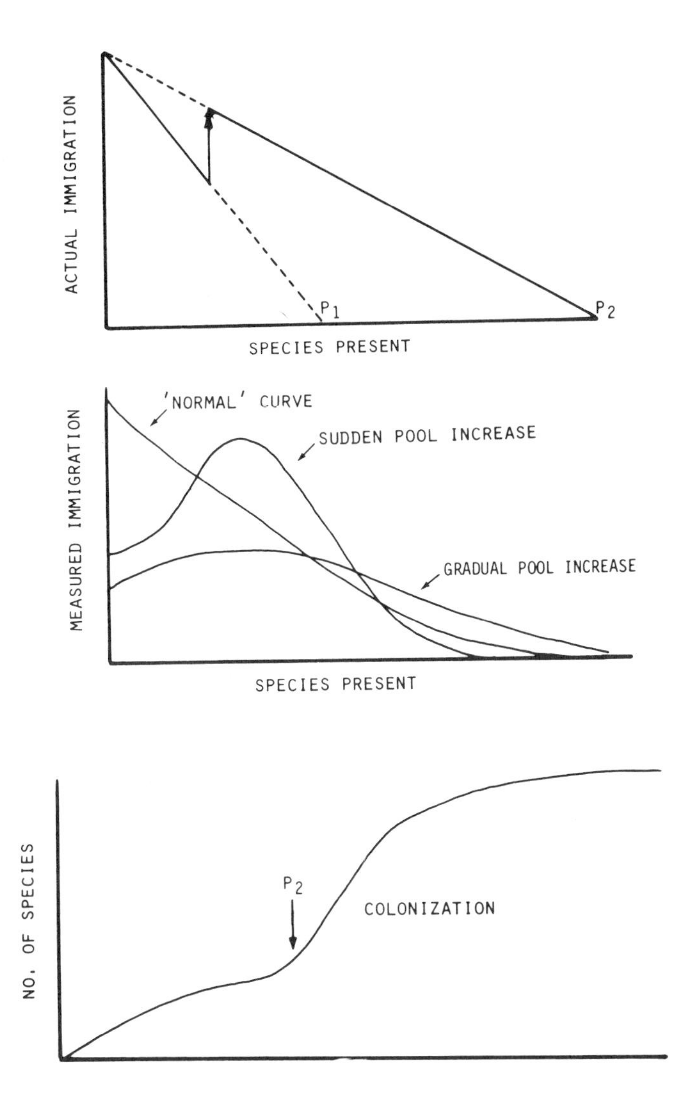

Figure 10. The effect of an increase in the size of the species pool on immigration and colonization. Top: A hypothetical increase in the species pool from P_1 to P_2 and the change in the actual immigration rate curve. Center: The possible effects a change in pool size might have on the *measured* immigration rate. The 'normal' curve shows a typical immigration rate curve. Different types of immigration curves can result, depending on whether the pool size changes suddenly (as in the top figure) or gradually. Bottom: The expected colonization curve if the pool size increased to P_2 at the time indicated by the arrow.

11). Substrates first must be conditioned before many species will attach; these species are therefore excluded from the pool. Secondly, as in most communities there are species that require the presence of certain other species before they will immigrate. For example, host species are necessary for mutuals, commensals and parasites to immigrate onto an island while prey species are necessary before predators can survive on an island. In this study, the effects of these influences on the pool size are probably small. Predators and herbivores were not included in any analyses. Fewer than 10 symbiotic species have been identified within the community, not enough to change the pool size *measurably*. Finally, and most importantly, as the flat panels are colonized by invertebrates, the physical structure of the panels is increased. This structure is created by the living organisms and remains as empty tests and vacated tubes when the animals die. These structures can collect sediment and debris, provide additional sites for firmer attachment, and create crevices and other cryptic areas that may protect individuals from predators and/or physical damage. As a result of this increase in structure, there will be a definite increase in the number of microhabitats on a panel, thereby increasing the potential for colonization by new species. The number of species with realistic probabilities of immigrating and, thus, the pool size, should therefore increase.

Successional changes in pool size and immigration rates are supported by several observations. First, over 40% of all species found on panels during the study (107 of approximately 250) were never found on panels that were exposed for less than two months. It may be possible to attribute part of this percentage to the initial absence of bacterial films.

Secondly, as shown earlier (Table III), the number of species present often has no apparent effect on the immigration rate. This makes little sense, since species already present on an island cannot immigrate. And, as species accumulate, fewer immigrant species remain, and the rate should decrease. As can be seen in Figure 11, this lack of a significant regression does not necessarily imply a lack of pattern. In the summer and fall panels, the absence of an inverse relationship between immigration and the number of species present results from very low immigration rates when few species are present. For the summer panels, there are 12 points showing low immigration rates with few species present. These immigration rates were measured on the three replicates during the first four sampling periods after initial exposure. Likewise, in the fall series, the three low values for immigration with zero species present were recorded in the first sampling period after initial exposure. If these points are excluded, there is a general decrease in immigration with increasing species number. These data are suggestive of the patterns illustrated in Figure 10 and are consistent with a hypothetical change in pool size. The

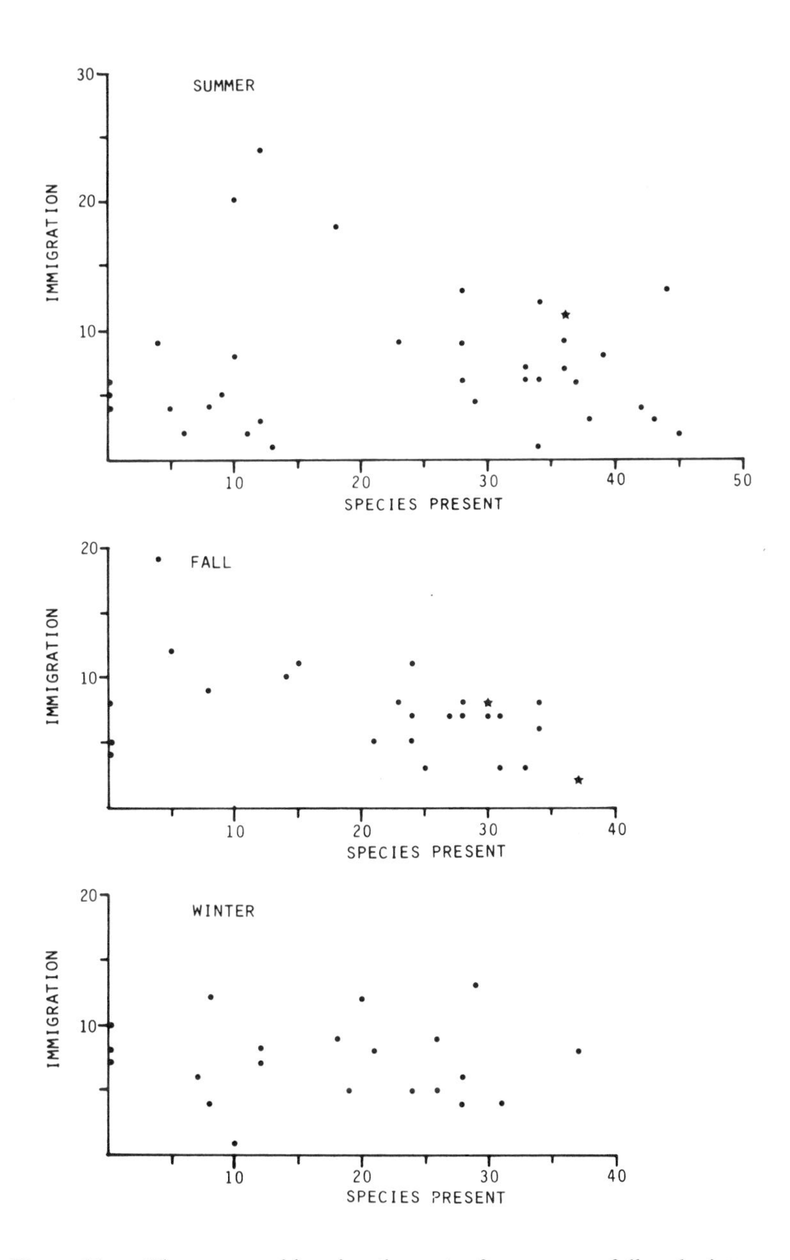

Figure 11. The measured immigration rates for summer, fall and winter series panels at the station inside the San Mateo kelp forest (see Figure 6 for the colonization curves). Immigration was measured as the number of new species on a panel per sampling period (about 6 weeks). Stars indicate two data points at the same coordinates.

immigration rates for the winter series show no apparent relationship to the number of species present. This absence of pattern may result from a gradual increase in the species pool (Figure 10); however, there are no data to test this hypothesis.

Finally, the colonization of the four seasonal series of substrates inside the San Mateo kelp forest (Figures 3, 6) suggests a structural effect. Two of these, the summer 1978 and spring 1979 series, at times accumulated more than 40 species per panel. One of the summer panels and all three of the spring series apparently equilibrated at about 45 species. As was discussed in an earlier section, the number of species on the remaining two summer panels decreased radically with the extreme dominance of the colonial polychaete, *Salmacina tribranchiata.* However, when the *Salmacina* colonies were removed by winter storms, thus exposing the underlying structure, the diversity rebounded to over 40 species. The remaining two series of substrates (fall and winter) never accumulated this many species. In fact, such high numbers of species were never found on any series at any other station. A probable explanation for the high diversity stems from the fact that the summer and spring San Mateo substrates alone were initially colonized by large numbers of barnacles (1000 individuals/panel). For the summer substrates this occurred in the fourth month after exposure (October 1978); for the spring panels, the barnacles recruited within the first two months (before the first sampling). These barnacles were preyed on by the flatworm *Stylochus tripartitus* and few individuals survived more than two months. However, the predators removed only the living tissue; the calcareous shells of the barnacles remained firmly cemented to the panels. These tests provided a great deal of structural relief in the form of hollow, truncated cones. On the summer 1978 panels (Figure 3) the numbers of species increased much more rapidly after the recruitment (and death) of the barnacles, while the colonization rate on the spring panels was initially higher than normal. Since these two series of panels were the only ones to both be exposed to such a single large recruitment of barnacles and to accumulate more than 40 species per panel, there is a strong suggestion that added structure leads to a higher equilibrium number of species.

This possible effect of structure was tested experimentally. Panels with at least 50% cover by barnacles were collected from a similar study near Santa Cruz Island. These substrates were put in a 10% chlorine bleach solution to remove all organic tissue. They were then rinsed thoroughly with water and exposed along with normal flat panels in both the San Mateo and San Onofre kelp forests in November of 1979. These panels were then sampled in March of 1980. As can be seen in Table VII, at both stations the panels with barnacle structure had significantly more species than the flat panels.

Table VII. Comparison of Panels with and without Structure in the San Onofre Kelp Forest and in the San Mateo Kelp Forest

	San Onofre Kelp		San Mateo Kelp	
	Structure	**Flat**	**Structure**	**Flat**
Mean Number of Species	23.25	14.17	30.06	10.83
Standard Error	1.80	1.22	2.14	0.91
Number of Panels (N)	4	6	5	6
	$t=4.35$, $df=8$, $p<0.01$		$t=9.09$, $df=9$, $p<0.001$	

[a] All panels were exposed in November 1979 and analyzed in March 1980. Structure panels had at least 50% of their surface area covered by barnacle tests.

Dean [20] has demonstrated that artificial substrates with structure have higher rates of colonization than those without. In this study, I have found that increased structure not only leads to more species colonizing a panel but that these substrates apparently can attain and maintain a higher equilibrium number of species. Some of this increase undoubtedly results from the increase in available area provided by the added structures. However, area increases alone cannot explain the increased diversity. For example, I measured several individual tests of the dominant species of barnacle on the San Mateo panels. Using these measurements, I estimated proportional relationships between basal diameter, height and diameter at the truncated apex of these tests. Assuming that barnacle tests were approximated by truncated cones, I calculated the area that the lateral surfaces of the barnacle tests added to a panel. I then used point count data to estimate the extent of the panel covered by barnacles. Assuming that the basal area inside the barnacle was still available to immigrants, I calculated that the mean area of the summer panels was increased 2.01 times and the area of the spring panels 2.38 times over the area of the flat panels. Using the standard area-species equation $S=CA^{0.26}$, these areal increases should have resulted in the species number increasing by 1.20 and 1.25 times. By estimating the number of species present without structure as 30 (an average for other kelp forest series), increases in area of 4–6 times would be necessary for the panels to increase (1.4–1.6 times) to 42–48 species. The increases in the number of habitats or microhabitats that result from structural complexity must account for much of the additional increase. Artificial substrates are probably not unique in this respect. Certainly as a terrestrial island develops from barren land to grassland to shrubs to forest, both the structural complexity and habitable area (e.g., tree trunks) increase. Obviously, many species are restricted to these different habitats and can successfully immigrate only if a specific habitat is there. An increase in structure should therefore increase both pool size and the rate of immigration.

CONCLUSIONS

Since its publication, the MacArthur-Wilson theory of island biogeography has been the focus of a substantial number of investigations. Only a few of these studies [38,46,47] have questioned its validity; for the most part, it has become an accepted addition to ecological theory. Recently, Gilbert [38] has seriously questioned many of the data that have been used to support the theory. He has called attention to the deceptive simplicity of the model and its often uncritical acceptance. He gives the following as conditions "in order to demonstrate unequivocally that the MacArthur-Wilson model applies in any particular situation: (i) a close relationship exists between insular areas and the numbers of species they contain; (ii) the number of species remains constant; (iii) an appreciable fraction of the complement of species changes in identity over periods of time related to the scale of the system." I feel that the data and analyses presented above meet these criteria and demonstrate that the theory is applicable to this community. Furthermore, the results of similar studies conducted in different habitats [31,33,34] indicate a general applicability to sessile invertebrate communities.

As Gilbert [38] has shown, survey data and multiple regression techniques cannot truly test the MacArthur-Wilson model. The key parts of the model are its dynamics and not the area-species relationship that it predicts. The importance of the dynamic balance between immigration and extinction and the effects of other phenomena on this balance can best be tested experimentally. The validity of this approach has been demonstrated amply by Simberloff [15,16], Simberloff and Abele [17] and Simberloff and Wilson [19]. As islands, artificial substrates provide the benefits of replication with good controls and the ability to verify assumptions regarding the manner in which immigration and extinction rates change. A plethora of studies have examined static distributions. A small number of studies, using experimentally defaunated islands or artificial substrates have measured colonization. However, few studies have attempted the measurement of the controlling rates of immigration and extinction. The preceding analyses indicate that these rates *can* be measured on artificial substrates. Through manipulations such as transferring a substrate to a new location, artificial substrates offer additional ways to test the equilibrium model.

Artificial substrates *are* islands to the sessile species that colonize them. There should be little difference in how a sessile invertebrate responds to a natural cobble island or an experimental panel island. Likewise, any of the processes that we observe on artificial substrates should occur on other "natural" types of islands. Immigration should be affected by island size, isolation, pool size, the seasonal supply of propagules, and the existing species diversity of the island. Extinction should be affected by island size, diversity and, perhaps, isolation. Most importantly, the communities on artificial substrate islands are dynamic. Part of this dynamism can be accounted for by stochastic events as represented by the equilibrium model; another part will result from highly deterministic processes. For example, dominance by species such as *Salmacina* or *Mytilus* results in diversity decreases that cannot be predicted using an equilibrium model. The model treats all species equally and cannot account for changes associated with the presence of a particular species. On the other hand, complex seasonal fluctuations in the species diversity of islands can be predicted if it is known how the probabilities of immigration and/or extinction vary through time.

The equilibrium model thus has an important role in separating variation associated with changes in external conditions from the highly deterministic changes associated with the presence of particular species on an island. It allows us to form hypotheses that can be tested about the effects of island size, isolation, seasonality, etc. These hypotheses are nothing more than complex null hypotheses that we must attempt to reject. In the cases of dominance by *Salmacina* or *Mytilus* or the countering effect of the bryozoan mutual *Zanclea* on *Salmacina* dominance, the equilibrium model cannot predict the observed changes in diversity. However, in rejecting an equilibrium model we are led to other questions regarding the biological interactions between particular species and their effects on the island community.

Artificial substrates are experimental islands that are as dynamic as other islands. However, in addition, they can be replicated, manipulated with controls, and their colonization history controlled and/or monitored. These advantages allow us to explore the dynamics of species on islands (or in patches) in great detail. This can be done with a minimum of untestable assumptions and without inferring causes from observed pattern. Without qualification, theories of island ecology can be tested strongly and easily with experimental substrates. In this sense, they are excellent ecological islands.

ACKNOWLEDGMENTS

I would like to thank J. Haugsness, J. Deacon, C. Mann and R. Day for their assistance in the field and laboratory. J. Dixon, T. Dean, J.

Wright, J. Haugsness, R. Day and an anonymous reviewer offered valuable comments on earlier versions of this paper. The research was funded by grants from the Marine Review Committee of the California Coastal Commission.

REFERENCES

1. MacArthur, R. H., and E. O. Wilson. "An Equilibrium Theory of Insular Zoogeography," *Evolution* 17:373–387 (1963).
2. MacArthur, R. H., and E. O. Wilson. *The Theory of Island Biogeography* (Princeton, NJ: Princeton University Press, 1967).
3. Diamond, J. M. "Avifaunal Equilibria and Species Turnover on the Channel Islands of California," *Proc. Nat. Acad. Sci. U.S.* 68:57–63 (1969).
4. Diamond, J. M. "Assembly of Species Communities," in *Ecology and Evolution of Communities,* M. L. Cody and J. M. Diamond, Eds. (Cambridge, MA: Belknap Press, 1975), pp. 342–444.
5. Lassen, H. H. "The Diversity of Freshwater Snails in View of the Equilibrium Theory of Island Biogeography," *Oecologia* 19:1–8 (1975).
6. Power, D. M. "Numbers of Bird Species on the California Islands," *Evolution* 26:451–463 (1972).
7. Wilcox, B. A. "Supersaturated Island Faunas: a Species-Age Relationship for Lizards on Post-Pleistocene Land-Bridge Islands," *Science* 199:996–998 (1978).
8. Simberloff, D. S. "Equilibrium Theory of Island Biogeography and Ecology," *Ann. Rev. Ecol. Syst.* 5:161–179 (1974).
9. Smith, A. T. "The Distribution and Dispersal of Pikas: Consequences of Insular Population Structure," *Ecology* 55:1112–1119 (1974).
10. Smith, A. T. "Temporal Changes in Insular Populations of the Pika *(Ochotona princeps),*" *Ecology* 61:8–13 (1980).
11. Flessa, K. W. "Area, Continental Drift and Mammalian Diversity," *Paleobiology* 1:189–194 (1975).
12. Schopf, T. J. M. "Permo-Triassic Extinctions: Relation to Sea-Floor Spreading," *J. Geol.* 82:129–143 (1974).
13. Simberloff, D. S. "Permo-Triassic Extinctions: Effects of Area on Biotic Equilibrium," *J. Geol.* 82:267–274 (1974).
14. Brown, J. H., and A. Kodric-Brown. "Turnover Rates in Insular Biogeography: Effect of Immigration on Extinction," *Ecology* 58:445–449 (1977).
15. Simberloff, D. S. "Experimental Zoogeography of Islands: A Model for Insular Colonization," *Ecology* 50:296–314 (1969).
16. Simberloff, D. S. "Experimental Zoogeography of Islands: Effects of Island Size," *Ecology* 57:629–641 (1976).
17. Simberloff, D. S., and L. G. Abele. "Island Biogeography Theory and Conservation Practice," *Science* 191:285–286 (1976).
18. Simberloff, D. S., and E. O. Wilson. "Experimental Zoogeography of Islands: The Colonization of Empty Islands," *Ecology* 50:278–296 (1969).

19. Simberloff, D. S., and E. O. Wilson. "Experimental Zoogeography of Islands: A Two-Year Record of Colonization," *Ecology* 51:934–937 (1970).
20. Dean, T. A. "Structural Aspects of Sessile Invertebrates as Organizing Forces in an Estuarine Fouling Community," *J. Exp. Mar. Biol. Ecol.* 53:163–180 (1981).
21. Schoener, A. "Colonization Curves for Planar Marine Islands," *Ecology* 55:818–827 (1974).
22. Jackson, J. B. C. "Morphological Strategies of Sessile Animals," in *Biology and Systematics of Colonial Organisms,* G. Larwood and B. R. Rosen, Eds. (New York: Academic Press, Inc., 1979), pp. 499–555.
23. Scheltema, R. S. "The Dispersal of the Larvae of Shoalwater Benthic Invertebrate Species over Long Distances by Ocean Currents," in *Fourth European Marine Biology Symp.,* D. J. Crisp. Ed. (London: Cambridge University Press, 1971), pp. 7–28.
24. Harper, J. L. *Population Biology of Plants* (New York: Academic Press, Inc., 1977).
25. Jones, H. L., and J. M. Diamond. "Short-Time-Base Studies of Turnover in Breeding Bird Populations on the California Channel Islands," *Condor* 78:526–549 (1976).
26. Diamond, J. M., and R. M. May. "Species Turnover Rates on Island: Dependence on Census Interval," *Science* 197:266–270 (1977).
27. Cairns, J., Jr., M. Dahlberg, K. L. Dickson, N. Smith and W. T. Waller. "The Relationship of Freshwater Protozoan Communities to the MacArthur-Wilson Equilibrium Model," *Am. Nat.* 103:439–454 (1969).
28. Henebry, M. S., and J. Cairns, Jr. "The Effect of Island Size, Distance and Epicenter Maturity on Colonization in Freshwater Protozoan Communities," *Am. Mid. Nat.* 104:80–92 (1980).
29. Henebry, M. S., and J. Cairns, Jr. "The Effect of Source Pool Maturity on the Process of Island Colonization: An Experimental Approach with Protozoan Communities," *Oikos* 35:107–114 (1980).
30. Patrick, R. "The Effect of Invasion Rate, Species Pool, and Size of Area on the Structure of the Diatom Community," *Proc. Nat. Acad. Sci. U.S.* 58:1335–1342 (1967).
31. Dean, T. A. "Succession in a Marine Fouling Community: Changes in Community Structure and Mechanisms of Development," Ph.D. Thesis, University of Delaware, Newark, DE (1977).
32. Dean, T. A., and L. E. Hurd. "Development in an Estuarine Fouling Community: the Influence of Early Colonists on Later Arrivals," *Oecologia* 46:295–301 (1980).
33. Osman, R. W. "The Establishment and Development of a Marine Epifaunal Community," *Ecol. Monog.* 47:37–63 (1977).
34. Osman, R. W. "The Influence of Seasonality and Stability on the Species Equilibrium," *Ecology* 59:383–399 (1978).
35. Schoener, A. "Experimental Zoogeography: Colonization of Marine Mini-Islands," *Am. Nat.* 108:715–738 (1974).
36. Sutherland, J. P., and R. H. Karlson. "Development and Stability of the Fouling Community at Beaufort, North Carolina," *Ecol. Monog.* 47:425–446 (1977).
37. Osman, R. W., and J. A. Haugsness. "Mutualism Among Sessile Inverte-

brates: a Mediator of Competition and Predation," *Science* 211:846–848 (1981).
38. Gilbert, F. S. "The Equilibrium Theory of Island Biogeography: Fact or Fiction?" *J. Biogeog.* 7:209–235 (1980).
39. Simberloff, D. S. "Species Turnover and Equilibrium Island Biogeography," *Science* 194:572–578 (1976).
40. Cairns, J., Jr., and J. A. Ruthven. "Artificial Microhabitat Size and the Number of Colonizing Protozoan Species," *Trans. Am. Micro. Soc.* 89:100–109 (1970).
41. Jackson, J. B. C. "Habitat Area, Colonization, and Development of Epibenthic Community Structure," in *Biology of Benthic Organisms,* B. F. Keegan, P. O. Ceidigh and P. J. S. Boaden, Eds. (Elmsford, NY: Pergamon Press, Inc., 1977), pp. 349–358.
42. Li, J. C. R. *Statistical Inference, Vol. 1, A Non-mathematical Exposition of the Theory of Statistics* (Ann Arbor, MI: Edwards Brothers, Inc., 1964)
43. Thorson, G. "Light as an Ecological Factor in the Dispersal and Settlement of Larvae of Marine Bottom Invertebrates," *Ophelia* 1:167–208 (1964).
44. Connell, J. H., B. J. Mechalas and J. A. Mihursky. "Interim Report of the Marine Review Committee to the California Coastal Commission. Part II: Appendix of Technical Evidence in Support of the General Summary," Marine Review Committee Document 79-02(II) (1979).
45. Pielou, E. C. *Biogeography* (New York: John Wiley & Sons, Inc., 1979), pp. 176–177.
46. Lynch, J. F., and N. K. Johnson. "Turnover and Equilibria in Insular Avifaunas, with Special Reference to the California Channel Islands," *Condor* 76:370–384 (1974).
47. Johnson, N. K. "Origin and Differentiation of the Avifauna of the Channel Islands, California," *Condor* 74:295–315 (1972).

CHAPTER 4

THE VARIETY OF ARTIFICIAL SUBSTRATES USED FOR MICROFAUNA

Stuart S. Bamforth

Newcomb College
Tulane University
New Orleans, Louisiana 70118

INTRODUCTION

Microfauna consists of motile protists (protozoa), including chlorophyll-bearing flagellates that reflect the geochemistry of regions; heterotrophic protozoa that prey principally on bacteria; and micrometazoa that prey principally on the protozoa. The microfauna maintains the bacteria in log-growth phase, enabling bacteria to degrade organic matter more effectively. The diversity, abundance and species of protozoa provide information about bacterial activity, hence protozoa are integral members of the decomposer community. This relationship, together with rapid reproductive rate and their intimate contact with the environment, make them more useful monitors of aquatic environments than the more frequently studied macroinvertebrates and fish.

While this chapter emphasizes protozoa, the presence and nature of accompanying micrometazoa, ranging from rotifers and rhabdocoel worms to insect larvae and small snails, should be included in every report to make the information about the water being monitored more complete.

Protozoa associate in small assemblages or in microhabitats, which may be patches of plankton, clusters along the neuston (surface scum or microlayer of water), or submerged surfaces of inanimate objects or living organisms. Since populations of bacteria accumulate on or near surfaces, protozoa and small invertebrates also colonize surfaces and form an "Aufwuchs" community consisting of attached zooperiphyton (e.g., peritrichs, sessile rotifers) and a large number of vagile species.

Most vagile (swimming, gliding and crawling) protozoa are associated with substrates. Wilbert [1] found 73 of 130 species of motile fresh water ciliates in the Aufwuchs only, and another 14 vagile species in both Aufwuchs and plankton. Many zooflagellates associate with substrates, and the genus *Mastigamoeba* shows how amoebae have developed to exploit surfaces. Most heliozoa, despite their spherical morphology (adapting them to planktonic life), live primarily among vegetation. Many phytoflagellates swim in the vicinity of substrates.

Protozoa are dispersed passively through waters onto substrates, but their ability to colonize depends on the nature of the substrate and organisms already present. Protozoan habitats are easily and, in lotic situations, frequently destroyed by floods, storms and movements of large animals. However, the continual release of protozoa from sources (species pools) in the aquatic environment enables rapid recolonization; and, in a few days, the protozoa can be used to describe the condition of an aquatic habitat.

A biologist can collect protozoa on artificial substrates inserted into the aquatic environment more easily than sampling natural substrates in the water. Artificial substrates are less selective than natural and can serve to compare communities of different waters or in different locations in the same stream or lake [2]. The entire substrate can be collected intact and transported to the laboratory, whereas natural substrates must be scraped, the surface material drawn up into tubes or bottles, and bits of material removed from the water. The size of the artificial substrate is known before sampling. The substrate is easily placed into and removed from the aquatic habitat, often enabling quantitative as well as qualitative estimations.

PRINCIPAL METHODS

The use of artificial substrates was introduced by Naumann [3] more than 60 years ago to collect iron bacteria and recommended by him to study microbial communities. Hentschel [4,5] introduced glass slides and a variety of other artificial substrates for both qualitative and quantitative determinations of Aufwuchs organisms.

A wide variety of materials have been employed, the most common being asbestos, asbestos cement (eternite), clay, concrete, sheet metals, wood, celluloid, glass, organic plastics and nylon netting. The last three materials are the most commonly used and may be grouped into four categories:

1. glass slides
2. plastic petri dishes and plastic cover slips
3. polyurethane foam (PF)
4. nylon reticular slides and grids

These artificial substrates are suspended in the water and are collected by detaching the substrates from their moorings and placing them into containers under water for transport to the laboratory for microscopical examination. All substrates used in a given study should be of the same type and size, thus allowing direct comparison of a study of succession in a water and comparison between waters.

Glass Slides

By far the most commonly used artificial substrate has been 2.5 × 7.5 cm glass slides, which have the advantage that they can be collected under water by placing them in a jar of water, then transported to the laboratory and placed on the stage of the microscope for direct observation.

Slides may be suspended in horizontal or vertical position in waters. Horizontally placed slides collect a true Aufwuchs community but also a large amount of seston (detritus, mud, decaying organisms). The underside will contain fewer organisms but sometimes favor special species such as *Zoothamnium*. Vertically placed slides are colonized more slowly, but the populations on both sides are similar [6]. In flowing waters the slides collect sediment rapidly. This condition may be retarded by placing a screen in front of the slides, as in a diatometer or by placing the slides in a perforated box to retard seston deposition.

Glass slides have been submerged in waters in a variety of ways, almost all of which are derived from, or similar to, those described by Hentschel [4,5], in a variety of habitats from ponds to large lakes and rivers [1,7–12]. These methods have been reviewed by Sladekova [6] and are summarized here.

In most cases, slides (or their containers) are attached to a string suspended vertically in the water. The lower end is tied to a rock and the upper end to a float or overhanging structure such as a bridge, pier or tree root. Early workers clipped glass slides to slate plates and wooden boards [5,10] hung by wire and rope in the water. Slides also have been attached by strings and wires [5,9]. Many workers [6,12] insert glass slides in slits made in rubber stoppers attached to lines vertically suspended in water. Although most workers insert slides in a vertical position, Sladekova has inserted slides alternatively in horizontal and vertical positions in her studies of reservoirs (Figure 1a).

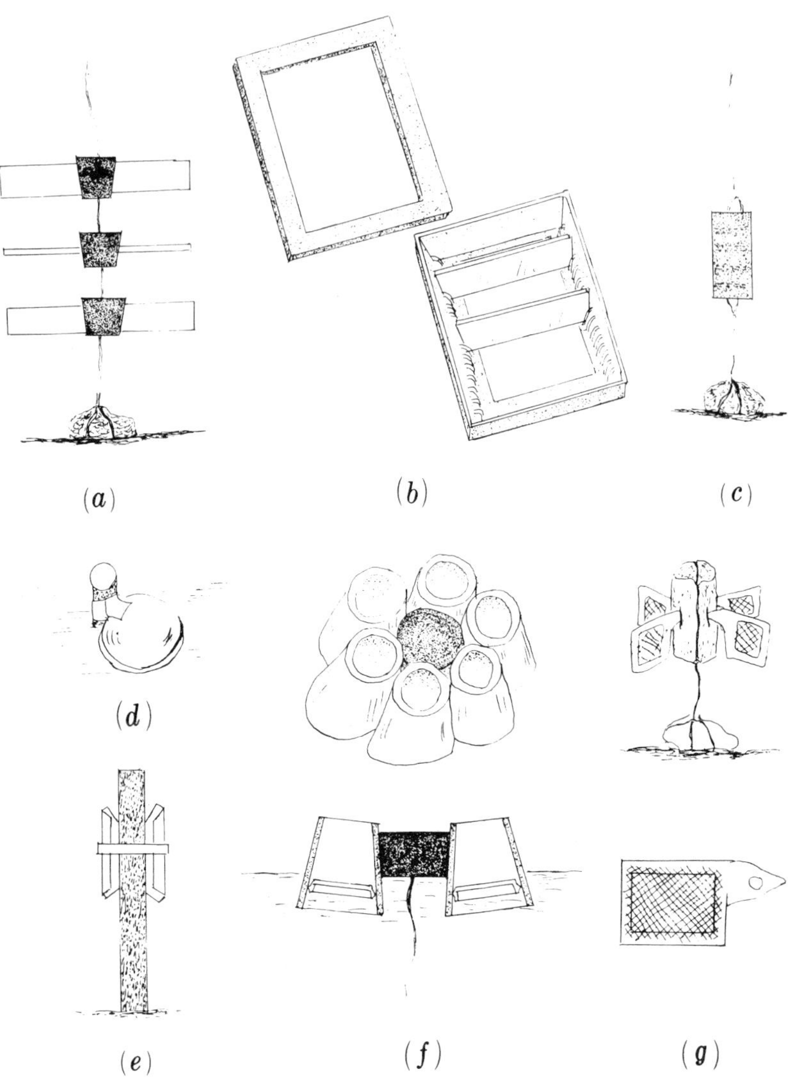

Figure 1. Applications of glass slide, plastic petri dish and reticulum grid methods: (a) suspended glass slides in alternate vertical and horizontal positions in rubber stoppers in reservoirs [6]; (b) glass slides in 8 × 11 cm plastic box; four slides can be placed in a box this size; (c) box of glass slides (in vertical position) suspended in water; (d) plastic petri dish taped to stake in water [13]; (e) two plastic petri dishes clamped to a stake a few cm above the bottom of a stream or pond; (f) top and side section views of carousel of six styrofoam cups (open at top) containing plastic petri dishes; (g) four reticulum grids inserted into wine cork attached to rock and detail of one grid.

Glass slides also may be placed in slide holders, which are suspended in water. Bissonette [14] made a rectangular wooden frame with 20 pairs of cuts to receive glass slides. The hinged lid and bottom holding the lids in place were made of zinc screen. Modifications have been made by several workers (e.g., Wilbert [1]). I have used 8 × 11 cm plastic slide boxes (Figure 1b,c) with most of the lid and bottom cut out to allow water to flow over the slides. Other modifications can be made easily. The slides should be placed at least 1.5 cm apart to allow colonization of organisms.

In lentic situations such as ponds, lakes and reservoirs, glass slides do not require as much protection against sedimentation. An interesting variation of the glass slide technique is "Treppenkultur" [5], in which a quantity of water is removed from the habitat and placed into a glass container or aquarium. Slides are placed in the container and collected at intervals of several days to assess organisms suspended in the water. Holm [10] supplemented a glass slide study of Suctoria in the Elbe River by placing 500-cm^3 samples of river water into aquaria and suspending slides in the aquaria to monitor swarmers (larval *Suctoria*).

The duration of exposure varies with the type of water sampled. In lotic environments, protozoan communities will build up in one to two days. In slow-flowing and lentic situations, a protist community matures in 2–4 weeks; in oligotrophic ponds and lakes up to 12 weeks [12]. Temperature is also a factor to be kept in mind.

In all glass slide techniques the slides are removed from their moorings under water and placed into jars of water for transport to the laboratory. One side of the slide is wiped clean and the slide examined directly under the microscope allowing the use of the 44X objective. The water in the jar in which the slide was transported is also examined as vagile species may leave or be jostled off the slides during transit. Nusch [12] found that 1% of peritrichs and 4–5% of vagile species were lost from the slide into the water in the jar.

Plastic Petri Dishes

A second method, introduced by Spoon and Burbanck [13], utilizes small 12 × 50 mm plastic petri dishes with tight-fitting lids. The plastic petri dish bottom is used to collect the organisms and may be placed into the water in two positions. The dish may be placed in an inverted horizontal position to retard collection of silt and attached to a stake (Figure 1d) or a flotation device. Spoon [15] has placed the inverted bottoms in open styrofoam cups attached to rubber stoppers, connected together with strapping tape (Figure 1f). The apparatus is tied either to a buoy in a river or to a below-surface line, and floats just below the surface of the water.

The second position is vertical, in which the plastic petri dishes are clamped to stakes or tubes driven into a substrate [16]. In this position plates collect organisms only a few cm above the sediment (Figure 1e). In the vertical position, colonization is rapid, and diverse populations build up in a few days.

In either horizontal or vertical position, collection consists of capping the dish with the watertight lid and transporting the dish to the lab. The outside of the lid and bottom are wiped to remove water and attached debris, and the entire dish is inverted and examined under the microscope. After scanning the bottom of the dish, the contents (13.2 ml) are emptied into a small container where the organisms are allowed to migrate to the top and bottom. These two zones are sampled to find species not detected in the initial examination.

The thickness of the plastic presents optical problems in determining identification of the organisms. Spoon [15] has met this problem by exposing plastic cover slips on the apparatus with the petri dishes and using the cover slips for oil immersion and scanning electron microscope studies. Earlier, Small and Ranganathan [17] utilized plastic cover slips mounted in cover slip holders to directly sample a polluted stream environment via scanning electron microscopy.

Polyurethane Foam

A third type of artificial substrate, introduced by Cairns et al. [18], is polyurethane foam (PF), which is a three-dimensional interstitial material, by contrast to the flat surfaces of the two preceding substrates.

Polyurethane foam is used commercially in cushions and pillows and as packing material, and may be purchased in blocks. Small, rectangular PF units measuring 5 × 6.5 × 7.5 cm are cut from the material [19]. Cairns and Ruthven [20] experimented with different sizes and found that this size sufficiently samples a natural population. Cairns and his colleagues, the principal workers using PF substrates, still vary their blocks from these dimensions [21], but generally this size of PF unit has been found satisfactory for routine monitoring of waters.

Twine is tied tightly around (almost bisecting) the middle of each PF unit and then tied to a nylon rope (Figure 2a). The units are spaced 30 cm apart along the rope to prevent entanglement. The rope is suspended horizontally in the water to vertical lines (Figure 2b). An alternate method in shallow waters is to attach a rock to a PF unit suspended from an overhead object, such as a tree limb or bridge (Figure 2c).

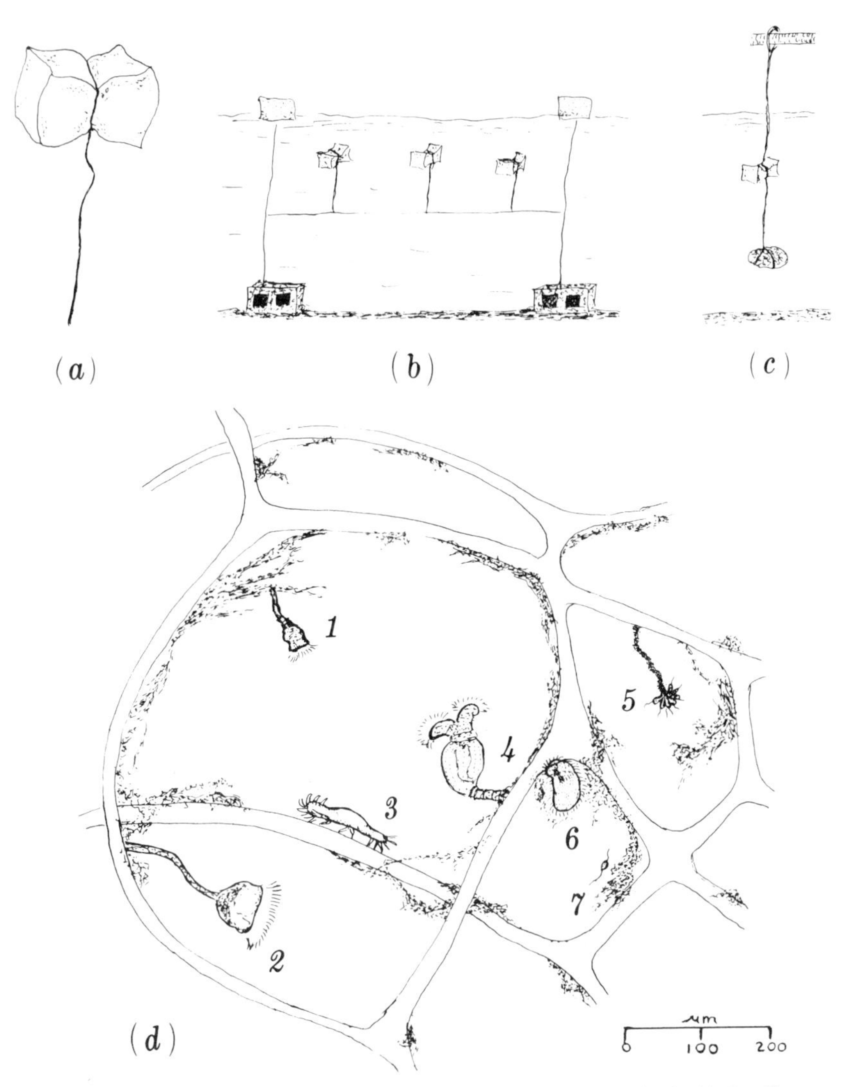

Figure 2. Applications of the polyurethane foam (PF) method: (a) detail of a PF unit showing the tight constriction in the center to prevent loosening of unit from line; (b) several units attached to horizontal line between two vertical lines [19]; (c) PF unit suspended from branch into stream; (d) composite diagram combining several observations of thin slices from a PF unit suspended in the Tchefuncte River below a sewage outfall for seven days and remaining overnight in the laboratory. Organic debris (very likely including bacterial colonies) is building on the fibers of the reticulum. 1, *Vorticella microstoma;* 2, *V. companula;* 3, hypotrich; 4, bdelloid rotifer; 5, *Anthophysa;* 6, *Trithigmostoma cucullus;* 7, Bodo.

In habitats with agitated or flowing water, such as large lakes or rivers, respectively, large protozoan populations can develop within a day [19,21]. In quieter waters, colonization is slower, requiring about five days.

To collect, a PF unit is detached from the rope and placed in a jar containing enough of the surrounding water so the unit is immersed. On arrival at the laboratory, the excess water is decanted and the water contents of the PF unit squeezed into the jar. The jar is placed in a position so that the organisms can respond to a single light source. The protozoa will migrate near the meniscus and on the bottom. Drops of water from these two positions are transferred to slides and can be analyzed fully under the microscope for species. Slides (usually about four) are prepared and examined until an asymptotic number of species is attained.

An alternate collecting method is to squeeze the PF units directly into the jars while in the field. This practice allows the PF units to remain attached to the lines and be used continuously in field studies.

Reticular Grids

A fourth method is to stretch nylon netting over a 16 × 25 mm rectangular hole in a plastic strip cut from a plastic slide into two halves, each with a pointed projection for insertion into a cork (Figure 1g). Wine corks are excellent holders and can contain four vertically positioned grids at right angles to each other. The cork is attached by twine to a weight for anchorage or may be tied to a submerged horizontal line (as for PF units). The cork floats, is resilient and holds grids tightly. Extensive colonization can occur within one day.

The grid may be collected in a small jar as for glass slides. If a detailed numerical study of the small grids is desired, however, the grid is collected in situ between two 5 × 7.5 cm glass slides, which have been taped at the edges and have the tape of one smeared with a sealant like Vaseline® or silicone lubricant. After the grid is collected between the slides, the assembly is aligned horizontally, removed from the water, and masking tape wrapped around it to keep the two slides together. The assembly is placed in a small box (e.g., slide box) for transport to the laboratory.

At the laboratory, a 5 × 7.5 cm glass slide is placed on the stage of the microscope and transported assembly placed on the slide. The grid is then examined for species and numbers per species. The tape and the top slide are removed carefully and the grid moved up and down in the pool of water on the lower slide to dislodge organisms attached to reticulum.

The top slide is replaced, and the assembly examined again for species not seen in the previous survey.

For a grid transported back to the laboratory in glass jars, a 5 × 7.5 cm glass slide is placed on the microscope stage and the grid placed on the slide. After surveying, the top slide is removed carefully and the grid moved up and down in the pool on the lower slide, removed and the top slide replaced. The contents are then examined for species not seen before. This rinsing procedure is necessary for older grids because silt accumulates on the reticulum, obscuring many species. The water in the collecting jar is also examined for any species not observed on the grid.

Reticular grids are most useful in very shallow habitats where the methods described on preceding pages are difficult to employ.

COMPARISON OF METHODS

Several users [15,16,19] of the plastic petri dish and PF methods have written of the advantages and disadvantages of the methods compared to glass slides. I conducted three series of experiments in September and October of 1980 to compare the monitoring abilities of the three principal methods.

Experimental

In each series, glass slides were placed in a 8 × 11 cm plastic slide box in which most of the areas of the lid and bottom were removed to allow maximum flow of water over the slides (Figure 1b). The boxes were suspended in the water (Figure 1c). Plastic petri dishes were placed in styrofoam cups arranged in carousels (Figure 1f) [15]. These and PF units either were attached to submerged lines and branches in the water or were suspended from overhanging branches with weights to keep the substrates in the water (Figure 2c). Each of the three types of substrates were collected on the same day and examined within 8 hours after collection.

The first series measured rate of colonization at a site near the Delta Regional Primate Center, Covington, Louisiana, in Abita Creek, a 30-meter-wide, relatively unpolluted stream flowing through rural countryside. Collections were made at 2, 7, 14 and, in one experiment, 21 days. Several reticulum grids and PF units were placed in a small slough to assess colonization of PF units in quiet water.

The second series compared the ability to monitor organic pollution with one site located 2 km above the Covington sewage disposal plant and a second site 50 meters below the plant effluent on the Tchefuncte River, a river larger than, but otherwise similar to, Abita Creek. Collections were made at 2- and 7-day intervals.

The third series measured low populations and siltation and were placed at sites in the Intracoastal Canal and Mississippi River in New Orleans. Vandalism restricted the sampling to a single site between the river levee and a fleet of anchored barges. Collections were made at 2-and 7-day intervals.

Examination of samples was conducted according to the procedures described earlier. For certain data analyses, protozoan species were grouped into three categories: (1) chlorophyll-bearing flagellates, (2) vagile heterotrophs, and (3) attached heterotrophs. All colorless flagellates, regardless of their taxonomic relationships, were placed with vagile heterotrophs. Thus, *Oikomonas, Monas* and colorless euglenoids like *Peranema* and *Anisonema* were grouped with zooflagelletes, *Sarcodina* and ciliates. Attached groups included peritrichs, *Suctoria, Stentor*, loricate hypotrichs and attached flagellates like *Anthophysa* and *Codonocladium.*

The Abita Creek series showed that 2–7 days of immersion time are sufficient for full colonization (in the absence of rainstorms); consequently, samples were collected at 2 and 7 days in subsequent studies. In the slough the PF units were colonized more slowly than planar substrates, but in lotic situations the colonization was as rapid as found by Henebry and Cairns [21].

The Tchefuncte survey showed two phenomena:

1. Although the three methods showed different numbers of species, all showed increased numbers of protozoa at the site below the sewage plant, reflecting the increased organic content of the effluent.
2. Following several heavy rains, the number of species below the sewage plant was reduced, probably due principally to dilution of the effluent caused by storm runoff from town streets.

The number of species at the upstream station increased due to land runoff and stirring up of the river and its tributaries. The upstream substrates contained more silt than before the storms, and more than the downstream station receiving the effluent.

On the Mississippi River collections, which contained much silt and low numbers as well as few species, all three methods recorded similar numbers of species.

In any collection, some species would be found on two or three of the substrates, but many would be found on one substrate only. (The Abita Creek study showed that 40–60% of the species on a PF unit were invaders since the previous collection.) For each collection, the total number of chlorophyll-bearing flagellates and vagile and attached heterotrophs were found by listing the species of these three groups from all three substrates to provide the total monitored for that collection. The number of species on each substrate was then compared to the total number to find the proportion of species collected by the substrate.

For chlorophyll-bearing flagellates and total numbers of protozoa, the three methods were comparable in the proportion of species collected. For vagile heterotrophs, the petri dish and PF methods usually would monitor between 50% and 75% of the total, whereas glass slides often monitored less than 50% (Figure 3).

The major collecting difference among the three methods was attached heterotrophs. The flat surfaces of the petri dish and the glass slide definitely favor attached species over the reticulum of the PF unit. The two planar methods monitored 50–100% of the attached species, whereas the PF unit often recorded less than 50% (Figure 3).

Dissection with a razor blade of PF units into thin strips showed that some attached species do enter the reticulum (Figure 2d) but that they are not always squeezed out. Species like *Stentor igneus, Vorticella convallaria, Campanella umbellaria* and *Epistylis plicatilis* were frequent but in low numbers on PF squeezings when appearing abundantly on planar substrates. *Vorticella companula, V. striata, Stentor muelleri* and *Pyxidium* sp. appeared rarely and in very low numbers in PF squeezings, while moderately to abundantly on planar substrates.

The planar substrates evidenced some selection. *Zoothamnium arbuscula* was most frequent and abundant on glass slides, but most attached species grew more abundantly on plastic petri dishes. *Pyxidium* sp. formed numerous rosette-shaped colonies, and Suctoria appeared most frequently and abundantly. Those petri dishes supporting massive numbers of peritrichs and *Stentor* spp. often contained large numbers of *Arcella vulgaris, Chilodonella* spp. and amphileptid carnivorous ciliates, especially *Trachelius*, a known predator on *Vorticella* [12].

A number of micrometazoa, which very likely compete with protozoa for bacteria and are known to prey on the protozoa, consequently appear in substrate samples and should be included in reporting. Bdelloid rotifers and gastrotrichs, almost universally present, were found frequently in all three methods. However, the petri dishes collected more frequently and greater numbers and species of monogont rotifers, ostracods, copepods, rhabdocoels, oligochaetes and dipteran larvae than glass slides or PF units.

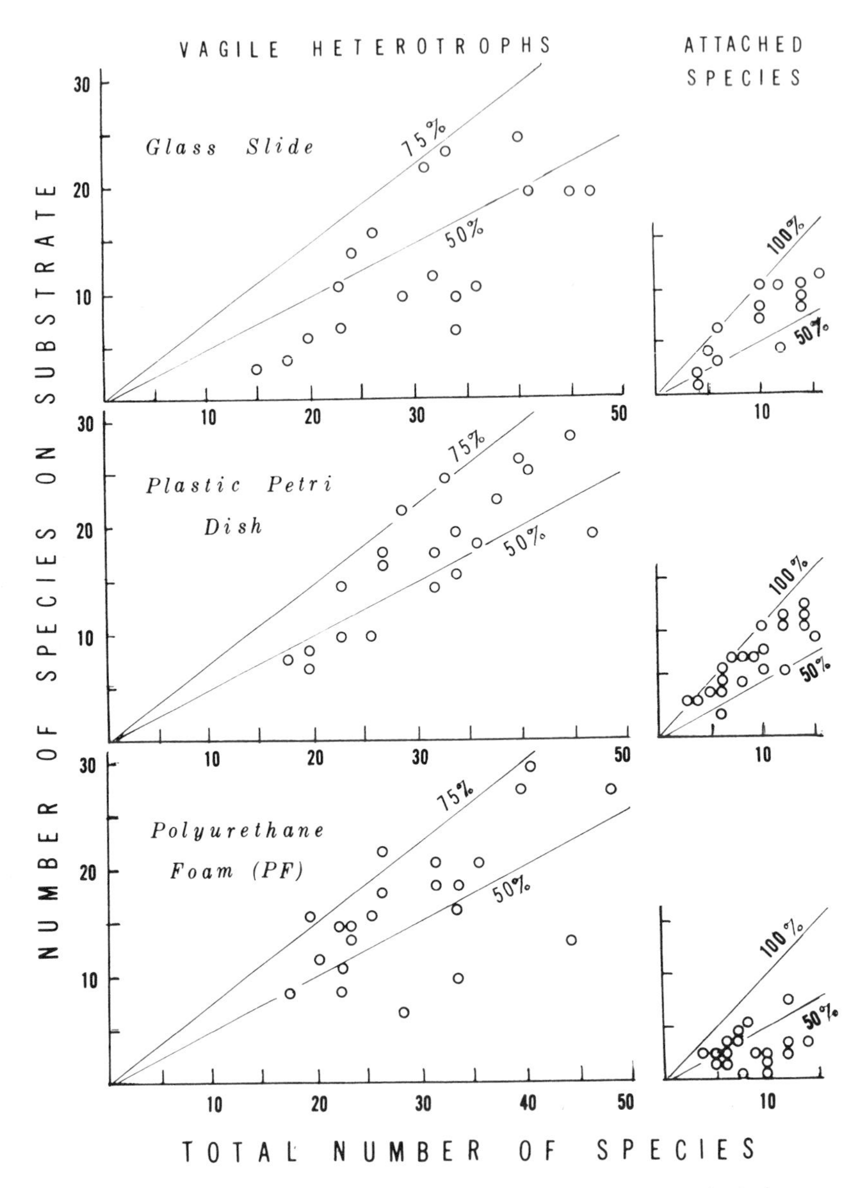

Figure 3. Monitoring ability of the three principal substrate methods for vagile heterotrophic and attached heterotrophic protozoa. The number of species colonizing a substrate is plotted against the total number of species monitored for the collection in the studies on Abita Creek and the Tchefuncte River. The glass slide method monitors fewer vagile heterotrophs, while the PF method measures fewer attached species.

Evaluation

Each of the three methods has its own advantages and disadvantages in regard to use in the environment, examination and organisms monitored.

The polyurethane (PF) method is the simplest. The units are easily cut from a large block of foam and do not require flotation. Collection and transport to the laboratory likewise are easy, and squeezings from the sample can be examined under the high power of the microscope. The method is least prone to vandalism because the buff-colored units are not very visible in waters and do not attract the attention of the more organized appearance of slide boxes or petri dish carousels.

The glass slide method is best carried out using slide boxes (especially in flowing waters). The slides are examined directly under the microscope and high power can be used.

The plastic petri dish method can employ stakes or, more effectively, a series of horizontally placed inverted dish bottoms in carousels. The area of the petri dish is equal to that of a glass slide. Transport to the lab is easy and without loss of organisms. The petri dish is examined on the microscope stage, but high power cannot be used. Also, the soft plastic scratches after several uses and these marks interfere with observation. The method is very prone to vandalism unless the carousels can be placed in safe locations, such as the buoys employed by Spoon [15] in the Potomac River.

Each method carries a bias. The PF units tend to dampen the number of attached species, whereas the two planar methods, especially the plastic petri dish, can emphasize attached forms and their associated vagile species. The PF units are not invaded as easily by micrometazoa as planar substrates, thus the protozoan community may be more complete. On the other hand, the small invertebrates often are important components in the microcommunity. The PF unit is easy and versatile, but in situations such as heavy organic pollution a planar method might be used to obtain supplementary data since attached protozoa and their metazoan competitors and predators can provide useful information about the condition of the environment.

REPORTING RESULTS

In addition to comparing the numbers of species at different stations in a body of water, the nature of the protozoan community also should be reported. Chlorophyll-bearing flagellates reflect the geochemistry of regions. For example, the presence of chrysomonads, dinoflagellates and *Gonyostomum* in Abita Creek and the Tchefuncte River reflect the low ion-containing waters of the Florida Parishes, whereas the euglenoid population of the Mississippi River reflects higher alkalinity and organic matter [22]. The restriction of chlorophyll-bearing flagellates to a few species of *Euglena* at the Tchefuncte site below the sewage disposal plant, together with the increase of many attached ciliates and the appearance of species such as *Urocentrum turbo* and *Paramecium caudatum*, attest to the organic enrichment of that site.

The relative abundance of species or groups provides information about the conditions of a habitat. Referring again to *Urocentrum* and *Paramecium*, a few individuals can occur in many situations, but large numbers indicate organic enrichment. Other quantitative features would be a paucity of chlorophyll-bearing flagellates, increased numbers of attached forms such as *Stentor* and peritrichs, and increased prominence of grazing protozoa like *Arcella, Trithigmostoma* and *Aspidisca*.

Nonprotozoa also should be reported. Diatoms and other algae often reflect the geochemistry of regions and, along with chlorophyll-bearing flagellates, are prey to many protozoa and micrometazoa and frequently attract herbivorous species of these heterotroph groups. Among the metazoa, an abundance of rhabocoel and oligochaete worms indicate organic enrichment, as is well known.

Cairns et al. [18] noted the relative abundance of species using a system of 1 to 5, and Spoon [15] used a rating scale ranging from 1 = rare to 7 = dominant (covering the entire substrate). These estimates are considered approximately arithmetic for large forms like *Stentor* and micrometazoa and exponential for tiny forms such as *Chlamydomonas* and zooflagellates. Assigning a number indicating relative abundance to each species found in a sample enables the worker to better describe his observations later. The scale should be small (I use 1–5) and may refer to numbers per microscope field, numbers per drop of sample on a slide (PF method) or estimates for the entire planar surface (in other methods). The worker may wish to report his findings in more generalized terms but he will have a quantitative basis from which to make this description.

REFERENCES

1. Wilbert, N. "Ökologische Untersuchungen der Aufwuchs und Planktonciliaten eines eutrophen Weihers," *Arch. Hydrobiol., Suppl.* 35: 411–518 (1969).

2. Cairns, J., Jr. "Zooperiphyton (especially protozoa) as Indicators of Water Quality," *Trans. Am. Micros. Soc.* 97: 44-49 (1978).
3. Naumann, E. "Eine einfache Methode zum Nachweis bezw. Einsammeln der Eisenbakterian," *Ber. Deutsch. Bot. Ges.,* 37: 76-78 (1919).
4. Hentschel, E. "Die Untersuchungen von Strömen," in *Abderhalden: Hand. Deutsch. Biol. Arbeitsmeth.* IX(2): 87-100 (1925).
5. Hentschel, E. "Abwasserbiologie," in *Abderhalden: Hand. Deutsch. Biol. Arbeitsmeth.* IX(2): 233-280 (1925).
6. Sladekova, A. "Limnological Investigation Methods for the Periphyton ("Aufwuchs") Community," *Bot. Rev.* 28: 286-350 (1962).
7. Agamaliev, F. G. "Ciliates of the Solid Surface Overgrowth of the Caspian Sea," *Acta Protozol.* 13: 53-83 (1974).
8. Fauré-Fremict, E. "Quelques résultats obtenus avec la méthode des lames immergées," *Bull. Soc. Zool. France* 56: 479-482 (1931).
9. Hammann, I. "Ökologische und biologische Untersuchwungen an Süsswasserperitrichen," *Arch. Hydrobiol.* 47: 177-228 (1952).
10. Holm, E. "Uber die Suctorien in der Elbe bei Hamburg und ihre Lebensbedingungen," *Arch. Hydrobiol., Suppl.* 4: 389-440 (1928).
11. Kralik, V. "Ein Beitrag zur Biologie von loricaten peritrichen Ziliaten, ins besondere von *Platycola truncata* Fromentel (1879)," *Arch. Protistenk.* 105: 201-258 (1961).
12. Nusch, E. A. "Ökologische und systematische Untersuchungen der Peritricha (Protozoa, ciliata) in Aufwuchs von Talspernen und Flusstauen mit verschiedenen Saprobitötograd (mit Model versuchen)," *Arch. Hydrobiol., Suppl.* 37: 243-386 (1970).
13. Spoon, D.M., and W. D. Burbanck. "A New Method for Collecting Sessile Ciliates in Plastic Petri Dishes with Tight Fitting Lids." *J. Protozool.* 14: 735-739 (1967).
14. Bissonnette, T. H. "A Method of Securing Marine Invertebrates," *Science* 71: 464-465 (1930).
15. Spoon, D. M. *Survey, Ecology, and Systematics of the Upper Potomac Estuary Biota: Aufwuchs Microfauna, Phase I,* Final Report, Water Resources Research Center, Washington Technical Institute, Washington, DC (1975), p. 117.
16. Kusters, E. "Ökologische und systematische Untersuchungen der Aufwuchsciliaten im Königshafen bei List/Sylt," *Arch. Hydrobiol., Suppl.* 45: 121-211 (1974).
17. Small, E. B., and V. S. Ronganathan. "The Direct Study of Polluted Stream Ciliated Protozoa via SEM," *Scanning Electron Microscopy* 1970: 283-293 (1970).
18. Cairns, J., Jr., M. L. Dahlberg, K. L. Smith and W. T. Waller. "The Relationship of Fresh-Water Protozoan Communities to the MacArthur-Wilson Equilibrium Model," *Am. Nat.* 103: 439-454 (1969).
19. Cairns, J., Jr., D. L. Kuhn and J. L. Plafkin. "Protozoan Colonization of Artificial Substrates," in *Methods and Measurements of Attached Microcommunities: A Review,* STP 690, R. L. Weitzel, Ed. (Philadelphia: American Society for Testing and Materials, 1979), pp. 34-54.
20. Cairns, J., and J. A. Ruthven. "Artificial Microhabitat Size and the Number of Colonizing Protozoan Species," *Trans. Am. Micros. Soc.* 89: 100-109 (1970).

21. Henebry, M. S., and J. Cairns, Jr. "Monitoring of Stream Pollution Using Protozoan Communities on Artificial Substrates." *Trans Am. Micros. Soc.* 99: 151–160 (1980).
22. Bamforth, S. S. "Limnetic Protozoa of Southeastern Louisiana," *Proc. LA Acad. Sci.* 26: 120–134 (1963).

CHAPTER 5

SUBSTRATE ANGLE AND PREDATION AS DETERMINANTS IN FOULING COMMUNITY SUCCESSION

Larry G. Harris and Katherine P. Irons
Zoology Department
University of New Hampshire
Durham, New Hampshire 03824

The purpose of this study was to test the hypothesis that substrate angle and predation are determinants in fouling community succession. The primary experiment involved arrays of 1/10-m^2 plexiglass panels set up in horizontal and vertical planes against and away from the wall of a wooden crib to facilitate and inhibit access by sea stars, crabs and wrasses. Two replicates of the experiments were established in summer and winter, respectively, to test for temporal effects on long-term sucession. The results to date of this continuing study clearly show both substrate angle and predation effects, but no seasonal effects. Siltation and predation have interacted to exclude most species from the upper horizontal surfaces, the exception being *Mytilus edulis* in predator-free systems. Lower horizontal surfaces have higher diversities and more species with upright growth forms than vertical surfaces. Mussels dominate most surfaces on predator-free panels, regardless of substrate angle. Predation effects have included keystone predation (on mussels), larval filter (by tunicates), and cropping of colonial species (hyroids and ectoprocts); in each case, the survival and distribution of species has been altered. Interphyletic competitive interactions have been documented and the results suggest that size, longevity and a living surface layer are important mechanisms enhancing the competitive abilities of encrusting colonial forms such as sponges.

INTRODUCTION

Marine fouling communities are assemblages of sessile suspension feeders, including sponges, hydroids, ectoprocts, bivalves, barnacles, tunicates and their associated fauna; they are typical inhabitants of vertical surfaces and cryptic habitats on natural hard substrates in the coastal zone. Fouling communities have received a great deal of attention [1-23] partly because of their conspicuous effect on man's activities in the marine environment, but also because of their ease of access and the suitability of many species for laboratory study.

Most studies on fouling communities to date have been descriptive in nature. These studies have identified successional sequences and the seasonal appearance of species seen on panels and have correlated this information with changes in such abiotic factors as temperature and salinity [11,23-25]. Few studies have described the biological factors affecting the establishment of fouling communities. In a detailed study of the fouling community in Beaufort, North Carolina, Sutherland [15,17,20] found that predation had little impact on the structure of fouling communities and that very few species were effective, long-term competitors for space. Sutherland hypothesized that fouling communities are characterized by a lack of organisms serving as competitive dominants. Studies by Boyd [1] and Fager [4] on West Coast fouling communities appear to substantiate this hypothesis. In the Gulf of Maine, short-term panel studies by Fuller [5] and Normandeau [25] suggest a similar lack of dominant space competitors.

However, there are studies of fouling communities that suggest that competitive domination can take place over time [13,14]. Among the organisms observed to dominate large amounts of space in fouling communities are sponges [6,26], hydroids [8,9,27], scyphistomae [1], anemones [28-30], mussels [14], ectoprocts [6,13,15-18,20], and tunicates [13].

Connell [31] presented a model for succession in communities that states that while several routes of development are possible, space freed by disturbance ultimately will be taken over by one of the competitive dominants present in that system. In a similar vein, Sutherland [15] suggested that in fouling communities several stable endpoints of community structure were possible, depending on the environmental conditions at the time that free space becomes available. Osman [13] found that one of several competitive dominants ultimately would take over on vertical rock faces and fouling panels if left in place long enough.

Jackson [6] synthesized the information available on fouling communities and proposed that (1) competition for space is the primary selective force in fouling communities, and (2) the encrusting, colonial growth form seen in sponges, ectoprocts and tunicates is a competitive strategy to take and hold space. In mature fouling communities on vertical and undercut surfaces, most primary space does appear to be occupied by organisms with a colonial and encrusting growth form.

Connell [32] has proposed that succession to competitive dominance occurs in all systems and that periodic disturbance, both abiotic and/or biotic, prevents succession from running its full course to dominance by a limited number of species; this occurs in communities characterized by high diversity, including tropical rain forests [33], Indopacific coral reef communities [34], deep-sea benthic communities [35] and horizontal rock substrates below about 30 meters in the Gulf of Maine and on the west coast of the United States [36,37]. In these environments disturbance may be in the form of intense predation, and dispersal of individual species within the community appears to be the rule.

The disturbance mechanisms suggested to maintain diversity in fouling communities are size of substrate [13], predation and storm damage [13], sloughing off due to mortality of seasonal species [15,18,20], and competitive webs [6,7,38].

Relatively little information is available on the role of predation as a disturbance factor in fouling communities. Sutherland [15], Jackson [6], and Keough and Butler [10] have suggested that predation is of minimal importance in their systems, yet Osman [13] suggested that predation may be a form of disturbance in fouling communities. Karlson [8,9] found that grazing by the sea urchin *Arbacia punctulata* was important in providing free space for the hydroid *Hydractinia echinata,* which could then hold acquired space for years. Peterson [39] found that crabs were important predators on pilings in Barnegat Bay, effectively excluding mussels from these communities. Russ [40] showed that fish predation may remove the young stages of certain tunicates and ectoprocts.

Observations of fouling communities on pilings and floats in the southern Gulf of Maine indicate that at least four species are capable of competitively dominating large areas of substrate [26,41]; these organisms are the sponge *Halichondria panicea,* the hydroid *Hydractinia echinata,* the anemone *Metridium senile*, and the blue mussel *Mytilus edulis*. *Mytilus edulis* is capable of dominating fouling communities [14] in a manner similar to that reported for *M. edulis* in the New England rocky intertidal [42] and for *M. californianus* in the west coast intertidal [43–46]. However, intense predation pressure by fish, crabs and starfish on young mussels on pilings tends to limit *M. edulis* to floats, while sponges and anemones dominate piling communities.

Horizontal substrates in the tropics and at depths below about 35 meters in temperate seas are dominated by sessile animals with colonial growth forms [6,37,47–49]. By contrast to vertical substrates (fouling communities), on horizontal surfaces the encrusting growth form is relatively rare, and most species grow up off the substrate in at least a mounding form. Even the sponge communities in the Antarctic show a strong tendency to upright growth forms on horizontal surfaces [50]. Detritus and sediments accumulate on upper surfaces but not on vertical or undercut surfaces. Grazing predators such as sea urchins, starfish, crabs, lobsters and fish tend to be found primarily on horizontal surfaces providing intensive predation pressure. This suggests that the angle of the substrate is a critical factor influencing the growth form in colonial animals. An encrusting strategy is most effective on vertical surfaces and the ceilings of undercuts, where competition for space is critical and burial by sediments and disturbances from grazing predators is minimal.

These patterns of community organization seen on horizontal and vertical hard substrates are consistent in the majority of subtidal marine environments from polar seas to the equator where communities are dominated by sessile invertebrates.

We propose the following synthesis of the ideas discussed in the preceding section:

Fouling communities as they have been studied historically are limited to vertical and undercut surfaces. On these surfaces competition is the most important biological selective force. Natural selection favors aggregation or encrusting colonial growth as competitive strategies for long-lived species adapted to occupy primary space. The role of predation on vertical surfaces is relatively less important and will function to influence the successional sequence.

On horizontal hard substrates, predation pressure replaces competition as the primary biological selective force affecting sessile species. Sediment accumulation interacts with predation to provide selective pressure for mechanisms that favor escape from smothering and grazing predators such as upright growth forms, dispersal and large size.

Based on the previous work and the synthesis proposed we have a series of experiments underway to test the following hypotheses:

1. Given time, a successional sequence will take place in most communities toward domination of the majority of primary space by one or a few of the superior competitors present in the system being studied.
2. The angle of the substrate will influence the species composition and, therefore, community development and structure.
3. Accumulation of nonconsolidated particulate matter on upper horizontal surfaces will influence species composition and community structure by inhibiting the success of species with an encrusting, colonial growth form. Predation will influence which species will ultimately dominate upper surfaces.
4. Competitive interactions will be the mechanism by which primary space is dominated in fouling communities by vertical and undercut surfaces. Predation will influence which of several species wins.

In the following sections the results of some of these studies will be described and their implications discussed.

MATERIALS AND METHODS

The purpose of the continuing panel study was to test the effects of substrate angle and predation on succession on plexiglass panels. The panels are approximately 1/10 m^2 (40.5 cm $\times$ 25.5 cm) in size. The shape was designed to facilitate photographic sampling, while the size was chosen to reduce the likelihood that any species would dominate the entire surface quickly.

The research is being conducted under a cement pier at the Portsmouth Harbor Coast Guard Station on the Great Bay Estuary in New Castle, New Hampshire. This site is located at the mouth of the estuary where there is minimal variation in salinity. It is protected from winter storms, and access by divers is restricted. The site is accessible by car and shading from the pier minimizes algal recruitment. The variation in salinity since July 1979 has been 26 to 32 parts per thousand. Water temperatures have ranged from $-1°C$ to $18°C$.

There is a substantial body of fundamental information available on the ecology of the estuary. The environmental consulting firm Normandeau Associates has been conducting environmental impact studies on the estuary for more than eight years [25]. The monitoring has included hydrographic and fouling panel studies. There are also approximately 10 years of data on nutrient chemistry, chlorophyll, temperature and salinity available for a sampling station adjacent to the pier [51,52]. The fouling fauna of the Great Bay Estuary is well documented [25,53] and information on fouling community structure at the mouth of the estuary also is available.

The experimental setup and sampling regime of the panel study is outlined in Table I and Figure 1. The panels are arrayed in strings with the long axis of each panel parallel to the current at approximately 15 feet below mean low water. The predator access panels are attached by ropes to a wall of large wooden beams, which is part of a boulder-filled crib supporting the pilings to the pier. This combination of beams and large rocks provides cover for high densities of crabs, lobsters, fishes and sea stars, which readily forage over the panels adjacent to the beams. The vertical panels are flush against the beams, while the horizontal panels are arrayed perpendicular to the wall with one edge touching a beam. The predator-free panels are suspended away from the beams and are essentially free of crab predation, although settlement of sea stars has occurred; these are removed from the panels regularly. The wrasses are reluctant to forage far from cover and only limited numbers have been observed hunting on the panels. Half of the predator access panels and half of the suspended panels were originally caged with sideless cages to

Table I. A Summary of the (I) Experimental Treatments and (II) Sampling Regimes Being Used[a]

I. Treatments

Predator Access	Angle of Substrate		
	Vertical	Upper Surface Horizontal	Lower Surface Horizontal
A. All predators (one edge against piling)	8 surfaces	8 surfaces	8 surfaces
B. Crabs and starfish excluded and fish reduced (suspended away from pilings)	8 surfaces	8 surfaces	8 surfaces

II. Sampling Regime

A. Photograph of total surface with 55 mm wide angle lens.
B. Photograph of delineated center (145 cm^2) with 3:1 closeup extension tube and framer.
C. Photographs of competitive encounters using 1:1 closeup lens and framer.
D. Documentation of predators seen and prey species where possible.
E. Collection by suction of unconsolidated sediments from three randomly selected points using a cofferdam to limit the area sampled.

[a] The substrates are rectangular plexiglass plates about 1/10 m^2 in surface area. The basic treatments have four replicates as listed. The initial set of panels was set up in July 1979. A replicate set was established in February 1980.

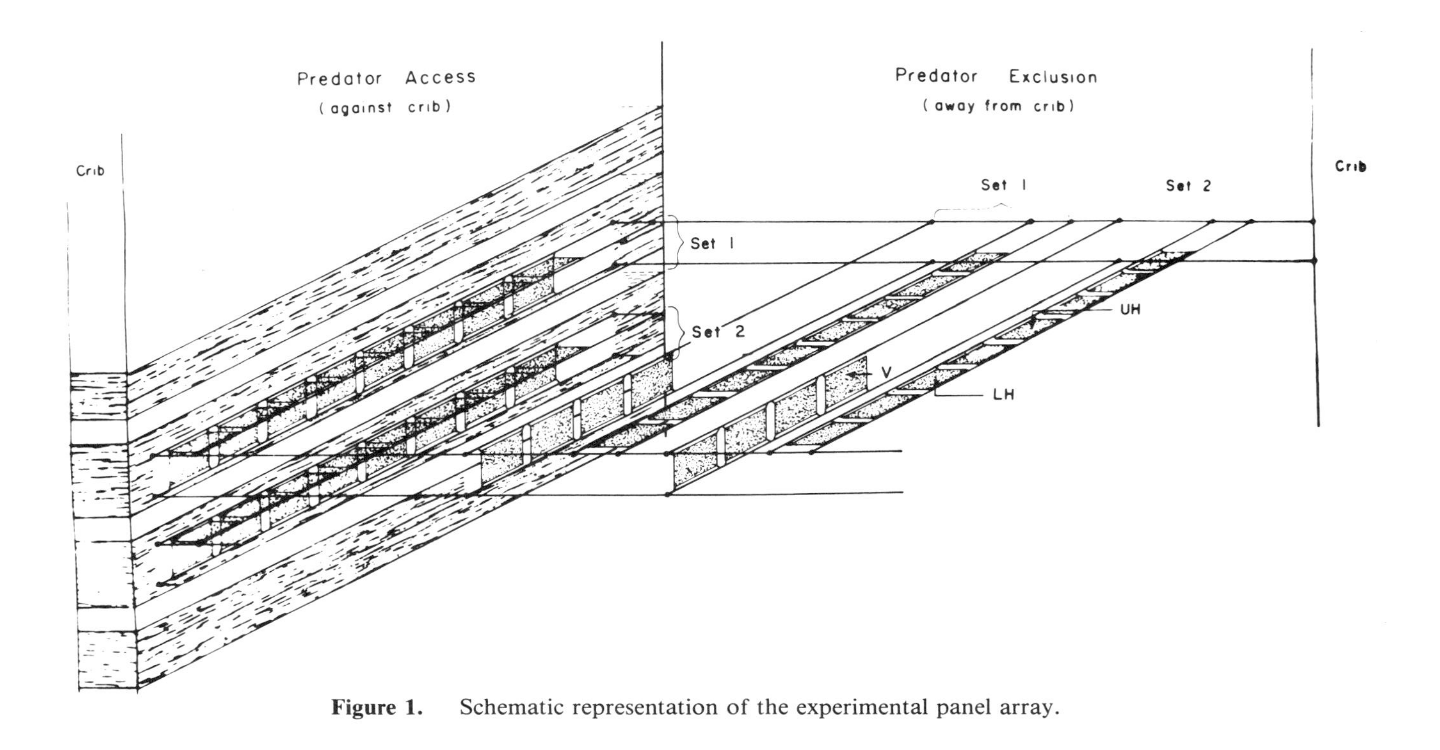

Figure 1. Schematic representation of the experimental panel array.

inhibit fish foraging. As there has been no measurable effect to date of caging on community succession, for this discussion the caged and cageless panels have been lumped to total eight replicates for each of six treatments (substrate angle—vertical, upper and lower horizontal—and predator access—yes, no).

Two identical sets of panels have been established to test for temporal effects on community succession. The first array was set up in early July of 1979 and a second set in February 1980, before the spring settlement of *Balanus balanoides*. The panels were sampled photographically at least once a month. Because color is an important indicator of species in many cases, color film and electronic flashes were used. Each entire panel (area 1/10 m^2) was photographed using a 15-mm wide angle lens on a Nikonos camera. To eliminate edge effects, the central portion of each panel, delineated by two brass screws, was photographed with a 3:1 Subsea extension tube and framer on a Nikonos; the effective area thus photographed is 145 cm^2. The resulting slides were projected on a grid of points (273 for the whole panels and 176 for the closeups) and percent cover estimated. A two-way analysis of variance was used to test the effects of substrate angle and predation.

Competitive interactions between dominant species on the panels were documented by photographing specific encounters with a 1:1 Subsea extension tube and framer on a Nikonos camera. This technique allows undisturbed sampling of competitive interactions. The resulting slides were then magnified and detailed observations made of interactions between species.

The feeding behavior of predators in the general piling community and on the experimental panels has been observed; 104 crabs and 46 wrasses also were collected and their stomach contents analyzed to determine the major food items being selected by these predators.

Both entire panel (15-mm lens) and closeup (3:1 extension tube) photographic slides from three sample sets were analyzed for this report. The slides from the November 1979 (after 4 months) and November 1980 (after 16 months) sample dates for the original set of panels and the November 1980 (after 9 months) sample date for the second panel array were used. This time period was chosen because species recruiting throughout the summer and early fall have grown to visible sizes and most winter mortality has not yet occurred. Use of these three data sets allowed us to compare time effects on a single panel set and also seasonal influences between periods of establishment on the two panel arrays. Closeup competitive interactions (1:1 extension tube) were recorded for a total of 216 slides, and results of encounters between certain dominant organisms (sponges, *Botryllus*, ectoprocts, *Balanus, Anomia* and *Metridium*) were analyzed.

RESULTS AND OBSERVATIONS

General Observations

The general community occupying the beams of the crib in the vicinity of the study site is dominated by encrusting sponges, hydroids, anemones and large barnacles. Soft corals, tunicates and encrusting and arborescent ectoprocts are interspersed throughout these assemblages and occasional small clumps of large *Mytilus edulis* are present. Decapod crustaceans (*Homarus americanus, Cancer irroratus, C. borealis* and *Carcinus maenus*) and the wrasse, *Tautogolabrus adspersus,* are the most common predators seen. The sea stars *Asterias vulgaris* and *Henricia sanguinolenta* are also abundant, but *A. vulgaris* at least is less common than in otherwise comparable habitats where the predatory crustacean population is lower; both lobsters and cancer crabs feed on *Asterias* spp. [37,54,55].

There is marked seasonality in activity patterns of the predators and recruitment of organisms into the system. Most of the fouling fauna recruit sometime between April and October, with exceptions including hydroids and mussels. The predators also show seasonal patterns. The wrasse *T. adspersus* disappears from the area in late October and does not return until May [56]. *Henricia, Cancer borealis, Carcinus* and *Homarus* also are absent in the winter months, while reduced populations of *Cancer irroratus* and *Asterias vulgaris* are still active and feeding.

Table II lists the majority of the species that have settled on the panels during the study to date. A conspicuous result of our observations is the low species diversity present on the panels, which is characteristic of the southern Gulf of Maine [25]. While all species of primary space occupiers collected have not been identified, it is evident that there are fewer than 40 and that the number of common species is less than 20. This is in stark contrast to the more than 300 species Jackson [6] has encountered on the undersides of coral plates on Jamaican reefs. Among the advantages of this relative simplicity is that it allows us to concentrate on interphyletic interactions [57] between representatives of most of the major taxonomic groups characteristic of fouling communities.

The characteristics given in Table II include relative size, external form, colonial or solitary, and surface covering. These factors all appear to be important in evaluating the adaptations of each species for avoiding predation and/or successfully competing for space. Table III and Figure 2 summarize the results of percent cover analyses for all species that occupied a mean of 5% on at least one substrate angle for one of the sampling periods included in this report. There are distinct dif-

Table II. List of Known Species Encountered on Primary Substrate of Fouling Panels (characteristics of size, morphology and surface type included)

Species	Occurrence	Size	Morphology	Colonial/Solitary	Surface Type
Porifera					
Haliclona loosanoffi	Rare	> 5 cm	Sheet	Colonial	Soft
Halichondria panicea	Common	>10 cm	Sheet	Colonial	Soft
Leucosolenia spp.	Rare	> 1 cm	Tubular crust	Colonial	Soft
Cnideria					
Campanularidae	Common	> 1 cm	Stolon	Colonial	Hard
Tubularia spp.	Rare	> 3 cm	Upright	Colonial	Hard
Metridium senile	Common	>10 cm	Mound	Solitary	Soft
Annelida					
Spirorbis spp.	Common	> 1 cm	Coiled tube	Solitary	Hard
Unidentified Terebellids	Rare	< 2 cm	Straight tube	Solitary	Soft
Mollusca					
Crepidula fornicata	Rare	> 2 cm	Flat	Solitary	Hard
Anomia spp.	Common	> 1 cm	Flat	Solitary	Hard
Mytilus edulis	Common	> 8 cm	Mound	Solitary	Hard
Hiatella arctica	Rare	> 1 cm	Mound	Solitary	Hard
Arthropoda					
Balanus spp.	Common	> 1 cm	Mound	Solitary	Hard
Ectoprocta					
Electra pilosa	Common	> 2 cm	Sheet	Colonial	Hard
Amphiblestrum flemingii	Common	> 2 cm	Sheet	Colonial	Hard
Bugula spp.	Common	> 1 cm	Upright	Colonial	Hard
Bowerbankia gracilis	Rare	> 1 cm	Stolon	Colonial	Hard

Chordata					
Aplidium spp.	Rare	> 5 cm	Mound	Colonial	Soft
Botryllus schlosseri	Common	>10 cm	Sheet	Colonial	Soft
Ciona intestinalis	Common	>10 cm	Upright	Solitary	Soft
Molgula spp.	Common	> 1 cm	Mound	Solitary	Soft
Didemnum candidum	Rare	> 2 cm	Sheet	Colonial	Soft

Table III. Percent Cover by Substrate Angle and Location for Each Common Space-Occupier Using Data Obtained from the Three Sets of Whole Panel Photographs

		% Cover					
		Access			Exclusion		
Space-Occupier	Data Set	UH	LH	V	UH	LH	V
Detritus	Ia	82	24	47	70	11	28
	Ib	75	16	59	49	10	10
	II	75	20	32	53	18	33
Hydroids	Ia	7	33	8	24	54	20
	Ib	6	16	4	1	2	1
	II	5	27	9	4	16	8
Botryllus	Ia	1	14	32	0	25	31
	Ib	0	8	5	0	5	3
	II	1	15	25	0	18	28
Mytilus	Ia	0	0	0	0	0	0
	Ib	0	1	0	47	55	74
	II	0	0	0	39	10	0
Barnacles	Ia	2	0	0	1	0	0
	Ib	8	12	0	0	3	1
	II	11	7	9	3	3	4
Anomia	Ia	3	10	2	1	5	4
	Ib	0	7	5	0	1	0
	II	4	8	8	1	3	3
Sponges	Ia	0	2	1	0	2	1
	Ib	2	25	6	0	18	11
	II	0	3	3	0	11	12
Spirorbis	Ia	4	8	4	1	5	3
	Ib	1	2	1	0	0	0
	II	1	5	6	0	5	4
Encrusting Ectoprocts	Ia	0	5	5	0	4	3
	Ib	0	1	1	0	0	0
	II	0	4	4	0	2	3
Upright Ectoprocts	Ia	0	1	0	0	1	0
	Ib	0	1	0	0	0	0
	II	0	4	1	0	1	2
Ciona	Ia	0	0	0	0	3	0
	Ib	0	1	0	0	4	0
	II	0	0	0	0	0	0
Molgula	Ia	0	1	0	1	1	0
	Ib	1	1	1	0	0	0
	II	1	2	0	0	2	1

Table III, continued

Metridium	Ia	0	0	0	0	0	0
	Ib	1	7	2	0	0	0
	II	0	1	1	0	0	0
Amphipod and Worm Tubes	Ia	0	0	0	0	0	0
	Ib	0	1	0	1	0	0
	II	0	0	0	1	2	1

[a] The complementary closeup photographs gave almost identical results: UH = upper horizontal; LH = lower horizontal; V = vertical; Ia = November 1979 for first panel set; Ib = November 1980 for first panel set; and II = November 1980 for second panel set.

ferences in the distribution patterns of most species, confirmed by two-way analysis of variance, that relate to time, substrate angle and location relative to the walls of the crib. Each species or group is covered below.

Detritus

The sediment load in the Great Bay Estuary is very high, and it was typical for all panel surfaces to be covered by a thin layer of fine particulate material, which had to be wafted away before the surfaces could be photographed. Therefore, detritus is also partially synonymous with free space since any organism occupying space below the removed silt would be recorded. Only on the upper horizontal surfaces was there any accumulation of particulates, and there were sharp differences in sediment type depending on the presence of mussels. In the absence of *M. edulis,* the slit was consolidated loosely and removed easily by wafting with a hand. The depth of this material on upper surfaces could reach a centimeter or more in less than two weeks, depending on the season, the amount of rain and whether crustaceans were walking over the surfaces regularly. Where mussels were present, the fine sediments were tightly consolidated feces and pseudofeces, which were difficult to remove even with strong hand waving. They also reached a depth of 3 cm or more in two weeks.

The upper surfaces had both the greatest detritus accumulation and free space. The predator-access vertical surfaces also had much free space, while the lower horizontal surfaces had the least free space.

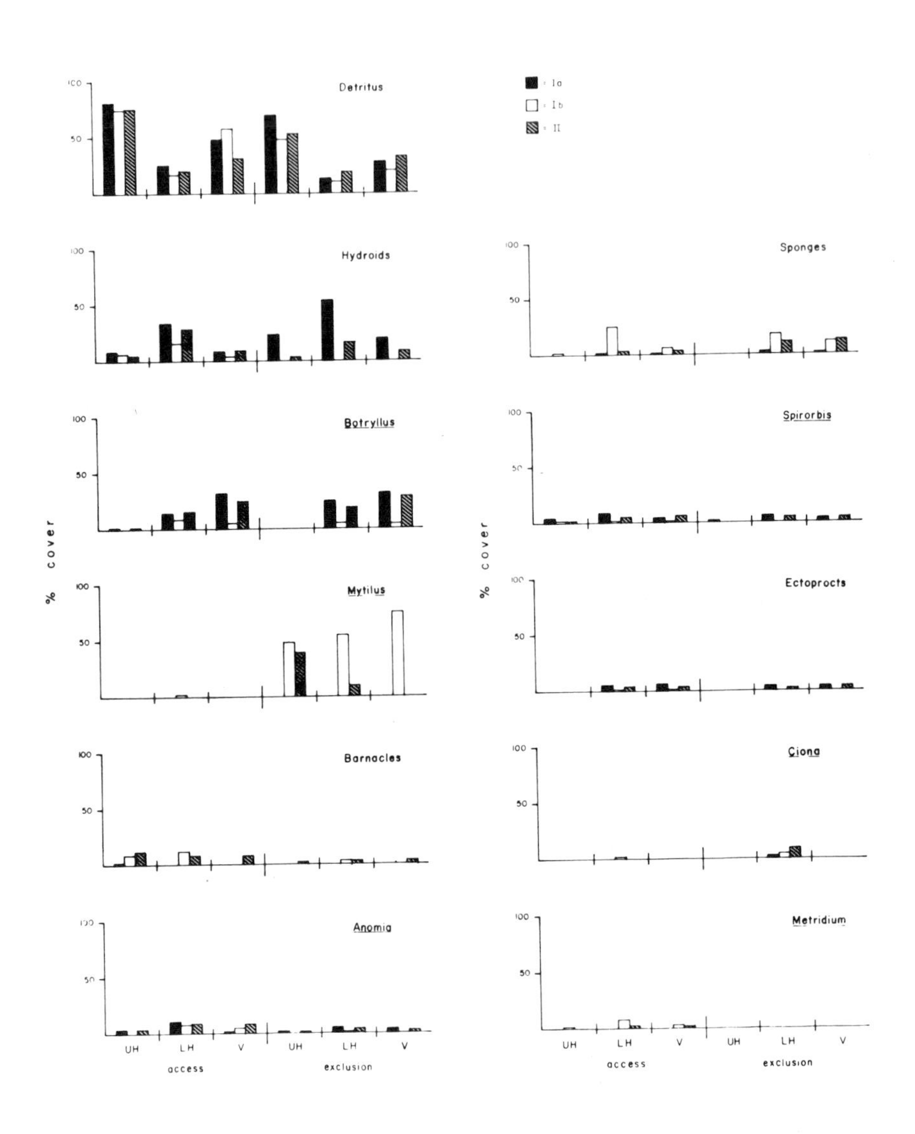

Figure 2. Summary of mean percent cover data for detritus and those species which had 5% or greater coverage on at least one substrate angle in one sampling period.

Hydroids

Figure 2 shows that hydroids were more prevalent on first-year panels (Ia and II) than on second-year panels (Ib) and that they appear to be most successful on those that have been in place the shortest time, in this case four months. Lower horizontal surfaces had the highest percent cover exclusive of location. There were differences in species between upper and lower horizontal surfaces, probably reflecting tolerances to sediment accumulation. The hydroids on upper surfaces were primarily *Obelia* species, which grew with long interstalk distances so that a loosely organized stolon mat covered much of the surface but was hardly discernible except on close examination. The hydroids occupying lower horizontal and vertical surfaces were dense colonies of *Tubularia* species that often obscured areas of primary substrate adjacent to the sites of attachment.

Mytilus edulis

Mytilus edulis is the dominant space-occupier of most of the second-year predator exclusion panels (Ib) and, to a lesser extent, on the second set of panels (II). Only a few small clumps of mussels are present on the lower horizontal surfaces of the predator-access panels, and then only on the second-year panels (Ib). Mussels were present in higher densities on the second set of panels (II), but there was almost total sloughing off of aggregations on lower horizontal and vertical surfaces in the late summer of 1980.

An interesting aspect of the mussel populations is the time of recruitment of the populations occupying the panels. Settlement occurred in mid- to late winter, as evidenced by the presence of mussels on panels established in February 1980. There is typically a heavy recruitment of *M. edulis* in summer in the Gulf of Maine, but there is a second settlement period in the winter, when predator populations are lowest.

The growth and domination of surfaces by mussels was very fast so that by mid-summer most of the predator exclusion surfaces were virtually a 100% cover of mussels. Then hummocking began to take place and sloughing off of aggregations occurred in late summer. On most of the second-year panels hummocking took place in the center of the panels. In the newly cleared patches were sponges, barnacles, *Anomia* and tunicates, which had obviously survived mussel overgrowth. In fact, the presence of these organisms may have been a partial cause of hummocking by occupying space needed for byssal attachment.

Anomia spp.

These flattened bivalves were most common on the predator-access panels, particularly the lower horizontal surfaces. Small *Anomia* spp. were present on all surfaces, but larger individuals ($\leq$1 cm) were relatively abundant only on the vertical and lower horizontal predator-access panel surfaces. The trend also appears to be toward decreasing survival over time.

Spirorbis spp.

These tiny serpulid polychaetes were conspicuous early settlers on all surfaces but decreased in density over time. They maintained the greatest populations on the predator-access panels and disappeared on panels dominated by mussels.

Metridium senile

This large anemone was found almost exclusively on the predator-access panels. A few large specimens crawled onto the panels from the adjacent beams, but most individuals were first observed as small (> 0.5 cm) individuals that had settled as planulae. There is a strong temporal component to their presence in that the highest percent cover is on the old set of surfaces (Ib).

The almost total absence of *M. senile* from the predator-exclusion panels suggests that the planulae of this species are demersal and ready to settle immediately on release from the parent anemone. While the density of anemones is highest on the cribs, there are numerous individuals on rocks below and up current from the panels that are suspended only 1 meter off the bottom and are arrayed parallel to current flow.

Balanus spp.

One of the primary reasons for selecting February as the time to establish the second set of panels (II) was to have them in place for the spring recruitment of *Balanus balanoides*. Indeed, there was for a limited time on all free surfaces almost 100% cover of this barnacle, but mortality was high and none of the barnacles observed in the November samples used for this study were *B. balanoides*. *Balanus balanus* and *B. eburneus* are most common on the predator-access panels and have survived best on the horizontal surfaces.

Sponges

Sponge colonies appeared in late summer and fall and grew through the winter. They have shown a high persistence in the presence of mussels and have increased in percent cover over time. The sponges present on upper horizontal surfaces are parts of colonies occupying lower horizontal surfaces that have grown over onto the top in areas where crab activity reduces silt accumulation.

The two dominant sponge species are *Halichondria panicea* and a *Haliclona* sp., both of which have encrusting growth forms. Small colonies of a *Leucosolenia* sp. were common as anastomosing tubular networks but did not occupy much space. Encrusting sponges are most prevalent on lower horizontal surfaces, although they are also conspicuous on the vertical surfaces of the predator-exclusion panels.

Encrusting Ectoprocts

The most numerous species was *Amphiblestrum flemingii,* followed by a few colonies of *Electra pilosa* and little else. These were limited to lower horizontal and vertical surfaces and did best during the first summer the panels were in place (Ia and II). The heavy silt load in the water at the study site may have decreased the success of this group in this study.

Ciona intestinalis

This large tunicate was found almost exclusively in one habitat—lower horizontal surfaces of predator-exclusion panels. They have shown high persistence in the presence of mussels and a number of specimens have lived for periods greater than a year.

Botryllus schlosseri

This encrusting, colonial tunicate was one of the most conspicuous species on lower horizontal and vertical surfaces during the late fall and early winter of each panel set, but they had a much reduced presence in the second year (compare Ia and II with Ib). Colonies were observed to undergo a regression, whether from senescence or disease is unclear, but this destruction appeared to be the primary cause of their decline on the panels.

Community Characteristics

Solitary versus Colonial

The species occupying primary space on the panels can be categorized as either solitary individuals or colonies. A further useful breakdown would be to identify those solitary species that actively aggregate through behavioral (i.e., *M. edulis*) and/or asexual reproductive (*M. senile*) mechanisms. If solitary and colonial forms are compared without regard for tendencies to aggregate, then the communities on the panels give the patterns seen in Table IVa and Figure 3. Colonial species dominate most surfaces except where mussels have covered the panels (Ib, UH, LH and V and II UH). The upper horizontal surfaces with colonial forms are either widely dispersed thecate hydroids or sponges growing up over the edge from below, compared to most surfaces where the colonial forms are crusts or dense bushes and tend to be more centrally located.

Aggregations of solitary species can dominate space in a manner similar to forms with encrusting colonial growth. If the two actively aggregating species, (*Mytilus* and *Metridium*), are added to the colonial forms, then almost all surfaces at all angles are dominated by dense aggregations or crusts (Table IVb and Figure 4). The one exception to this is on the upper horizontal, predator-access surfaces, where the major colonial form, thecate hydroids, consists of long stolons with occasional upright stalks; the same hydroid species on the pilings form dense bush-like colonies and occur primarily on vertical surfaces.

Encrusting versus Upright Growth Forms

Another approach to determining the community organization is to compare the relative importance of organisms with encrusting, sheet-like growth forms with those having upright and mounding shapes (Figure 5). Solitary and colonial species are found in both categories. In all data sets (Ia, Ib and II) lower horizontal surfaces have percent cover of upright forms that are almost equal to, or greater than, the cover by encrusting organisms (Table V). Mussels account for this large representation only on one set of lower horizontal surfaces (Ib, predator exclusion). *Mytilus* also accounts for the high percent cover of upright forms on Ib and II upper horizontal and Ib vertical predator-exclusion surfaces. Elsewhere, vertical surfaces are dominated by species with encrusting growth forms. The lower horizontal surfaces not dominated by mussels have complex aggregations of hydroids, *Metridium*, arborescent ectoprocts, barnacles

Table IVa. Percent Cover of Solitary and Colonial Organisms by Substrate Angle and Location Using Whole Panel Data Sets[a]

		% Cover					
		Access			Exclusion		
		UH	LH	V	UH	LH	V
Solitary Forms	Ia	9	19	6	3	14	7
	Ib	11	31	9	47	63	75
	II	17	23	24	42	32	12
Colonial Forms	Ia	8	55	46	24	86	55
	Ib	8	51	16	1	25	15
	II	6	53	42	4	48	53

[a]Solitary organisms include *Mytilus,* barnacles, *Anomia, Spirorbis, Ciona, Metridium;* colonial organisms are hydroids, *Botryllus,* sponges, encrusting ectoprocts, upright ectoprocts: UH = upper horizontal; LH = lower horizontal; V = vertical; Ia = November 1979 for first panel set; Ib = November 1980 for first panel set; II = November 1980 for second panel set.

Table IVb. Percent Cover of Solitary, and Colonial and Aggregative Organisms by Substrate Angle and Location Using Whole Panel Data Sets[a]

		% Cover					
		Access			Exclusion		
	Data Set	UH	LH	V	UH	LH	V
Solitary Organisms	Ia	9	19	6	3	14	7
	Ib	10	23	7	0	8	1
	II	17	22	23	3	22	12
Colonial and Aggregative Organisms	Ia	8	55	46	24	86	55
	Ib	9	59	18	48	80	89
	II	6	54	43	43	58	53

[a]Solitary organisms include barnacles, *Anomia, Spirorbis, Ciona, Molgula*; colonial and aggregate organisms are Hydroids, *Botryllus, Mytilus,* sponges, encrusting ectoprocts, upright ectoprocts, *Metridium.* Ia = November 1979 for first panel set; Ib = November 1980 for first panel set; II = November 1980 for second panel set.

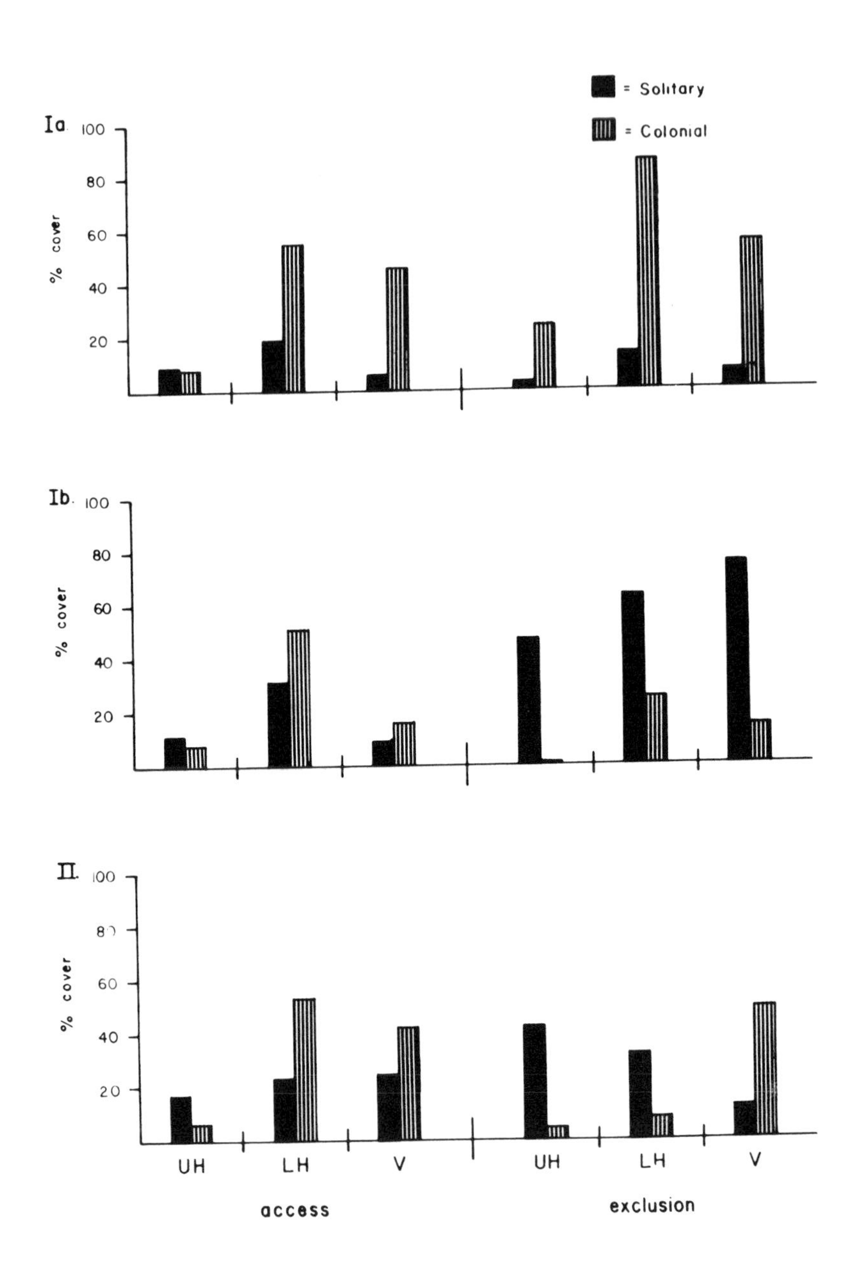

Figure 3. Comparison of accumulated mean percent cover for all solitary species versus those with colonial growth forms. The three sampling periods, substrate angles and predator access are all presented.

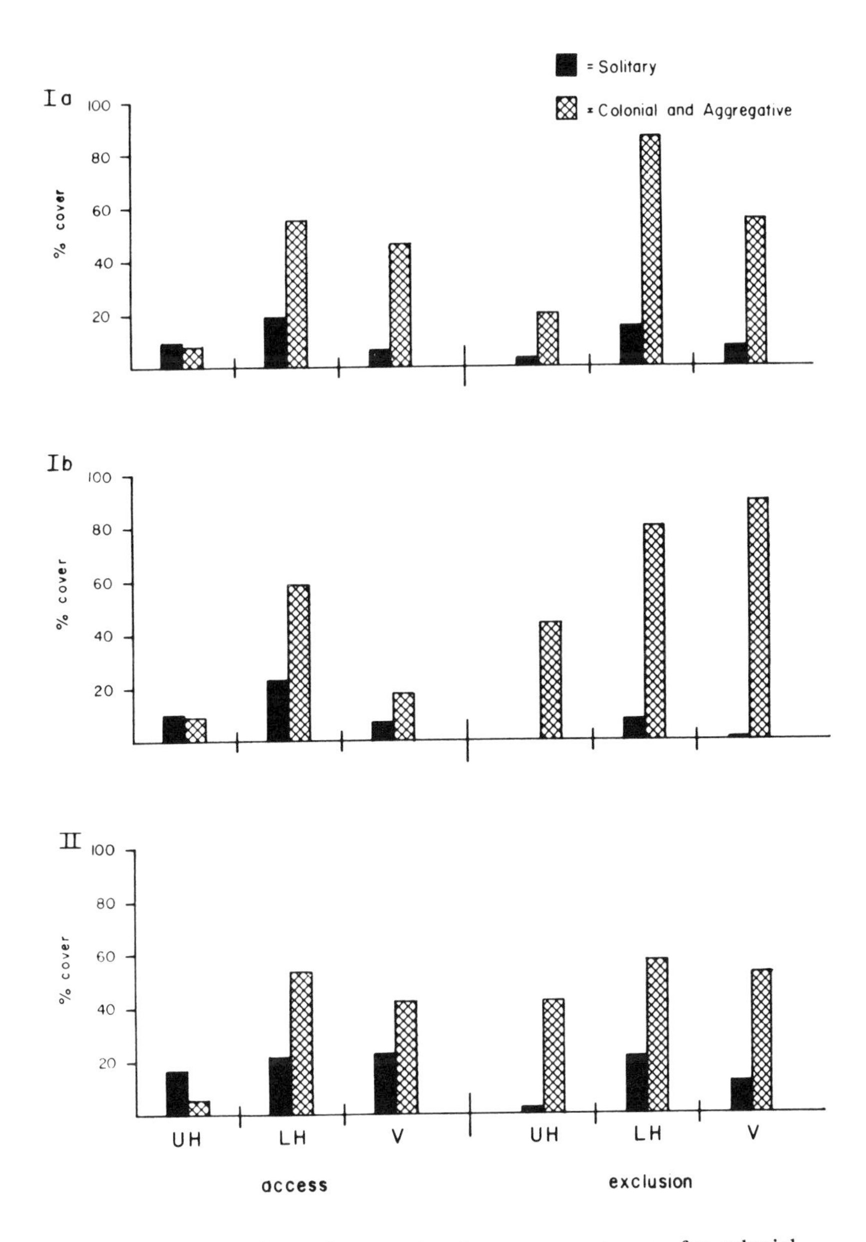

Figure 4. Comparison of accumulated mean percent cover for colonial species plus those solitary species that aggregate (*Mytilus edulis* and *Metridium senile*) versus the non-aggregating solitary species. The three sampling periods, substrate angles and predator access are all presented.

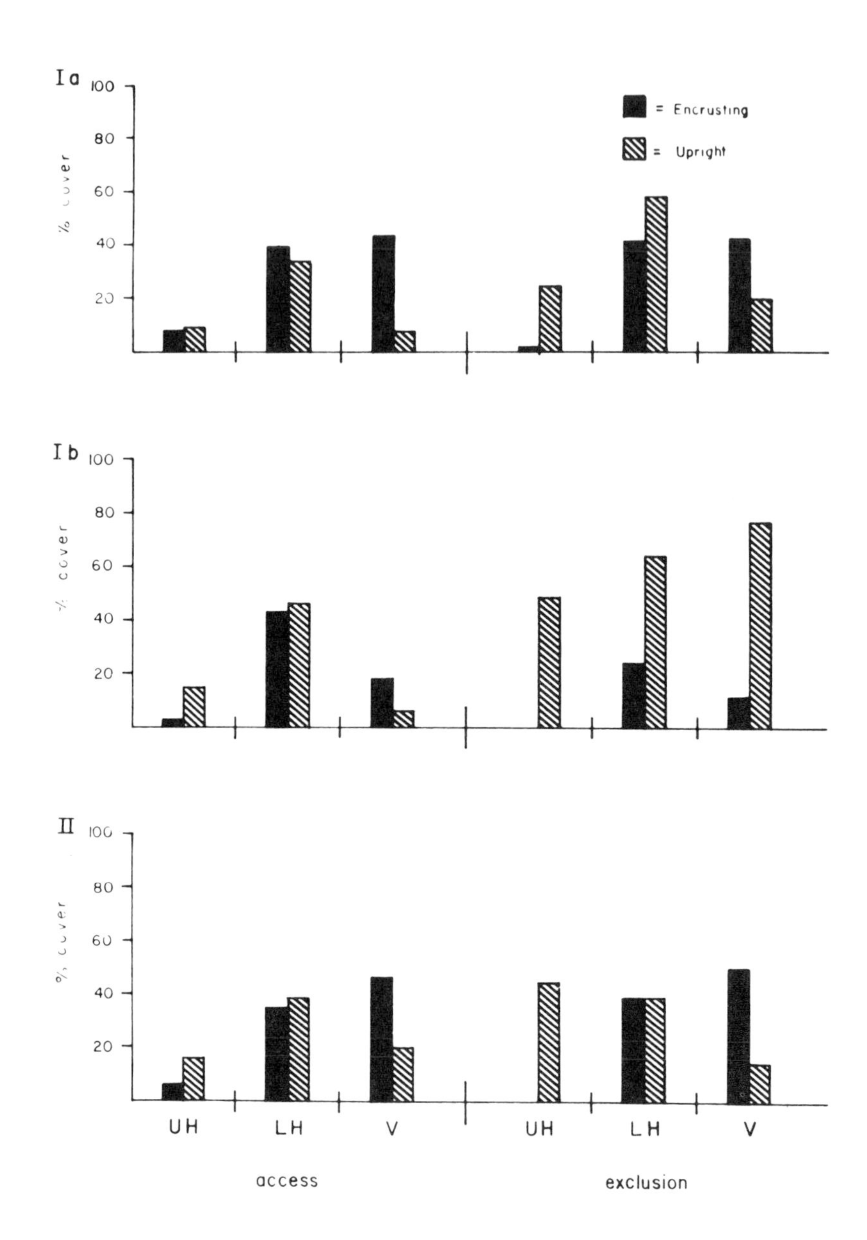

Figure 5. Comparison of accumulated mean percent cover for those species with encrusting growth forms versus those with upright and mounding morphologies. The data for each substrate angle and predator access for each sampling period is presented.

Table V. Percent Cover of Encrusting and Upright Growth Forms by Substrate Angle and Location Using Whole Panel Data Sets[a]

		% Cover					
		Access			Exclusion		
		UH	LH	V	UH	LH	V
Encrusting Forms	Ia	8	39	43	2	41	42
	Ib	3	43	18	0	24	11
	II	6	35	46	1	39	50
Upright Forms	Ia	9	34	8	25	58	20
	Ib	15	46	6	48	64	76
	II	16	30	20	45	39	14

[a]Encrusting forms include *Botryllus, Anomia,* sponges, *Spirorbis,* encrusting ectoprocts; upright forms include hydroids, *Mytilus,* barnacles, upright ectoprocts, *Ciona, Metridium.* UH = upper horizontal; LH = lower horizontal; V = vertical; Ia = November 1979 for first panel set; Ib = November 1980 for first panel set; and II = November 1980 for second panel set.

and *Ciona.* Hydroids were more common on lower horizontal surfaces during the first year (Ia and II), while anemones and tunicates increased in percent cover during the second year (Ib).

Predation

The role of predators in this experiment has taken several forms. The most obvious impact of predation has been the almost total exclusion of *Mytilus edulis* from the predator-access panels. All of the major predatory species observed were actively feeding on mussels (Table VI, Figure 6). Cunner had the highest percentage occurrence of mussels in their stomach contents (88%); *Cancer irroratus* was the second highest, with 80%. Wrasses and *C. irroratus* were the two most common predators observed foraging on the panels. *Asterias vulgaris* also actively feeds on mussels [36,41]. Cunner were observed foraging on the predator-exclusion panels, although not to the extent that they were feeding on the predator-access panels. It is possible that fish predation during the summer inhibited successful summer settlement of *M. edulis.* Mussels settling during the winter appeared to have grown to a refuge in size before cunner began active foraging in late spring.

The stomach content studies showed several differences between crabs and wrasses as predators on the panels. Wrasses had the highest percen-

Table VI. Summary of Stomach Content Analysis of Four Common Predators Showing Percentages of Stomachs Containing 13 Food Sources

Stomach Contents	*T. adspersus*	*C. irroratus*	*C. borealis*	*C. maenus*
Mussels	89	80	57	71
Barnacles	64	11	29	43
Hiatella arctica	51	7	0	0
Algae	47	22	29	14
Hydroids	21	1	0	21
Gastropods	17	1	0	7
Anomia simplex	13	1	0	0
Polychaetes	9	1	0	0
Amphipods	6	5	0	7
Ciona intestinalis	6	0	0	0
Strongylocentrotus droebachiensis	2	1	29	7
Mya arenaria	0	13	29	7
Ophiuroid	0	2	0	0
Empty	2	11	14	0
N	47	85	7	14

tages of barnacles, *Hiatella, Anomia* and *Ciona,* while the crabs often had *Mya* shells in their stomachs. The crabs actively forage in the sediments below the panels and while crabs were only collected up on the beams, it is obvious that at least, some, had been feeding on the bottom recently. *Asterias vulgaris* consistently feeds on *Balanus, Hiatella* and *Anomia,* and numerous empty shells of *Anomia* and *Balanus* on the predator-access panels attested to starfish predation on these two forms. The only predator found to eat tunicates was *T. adspersus,* which would explain the absence of *Ciona* from the predator-access panels.

Fish predation appeared to be more size-selective than for either crabs or starfish. Cunner were the only large predator regularly found associated with the predator-exclusion panels where mussels, *Ciona, Hiatella* and *Anomia* were common, but most were of large size.

The activities of crabs may have had two secondary effects on the predator-access panels. Crabs crawling around on the upper horizontal surfaces seemed to reduce the accumulation of detritus, thereby facilitating increased survival of sessile species that otherwise might be smothered. On vertical surfaces, the weight and disruption of crabs crawling over encrusting, colonial forms like sponges and *Botryllus* may have caused sufficient injury to inhibit growth and survival on that substrate angle because small colonies did appear but did not persist.

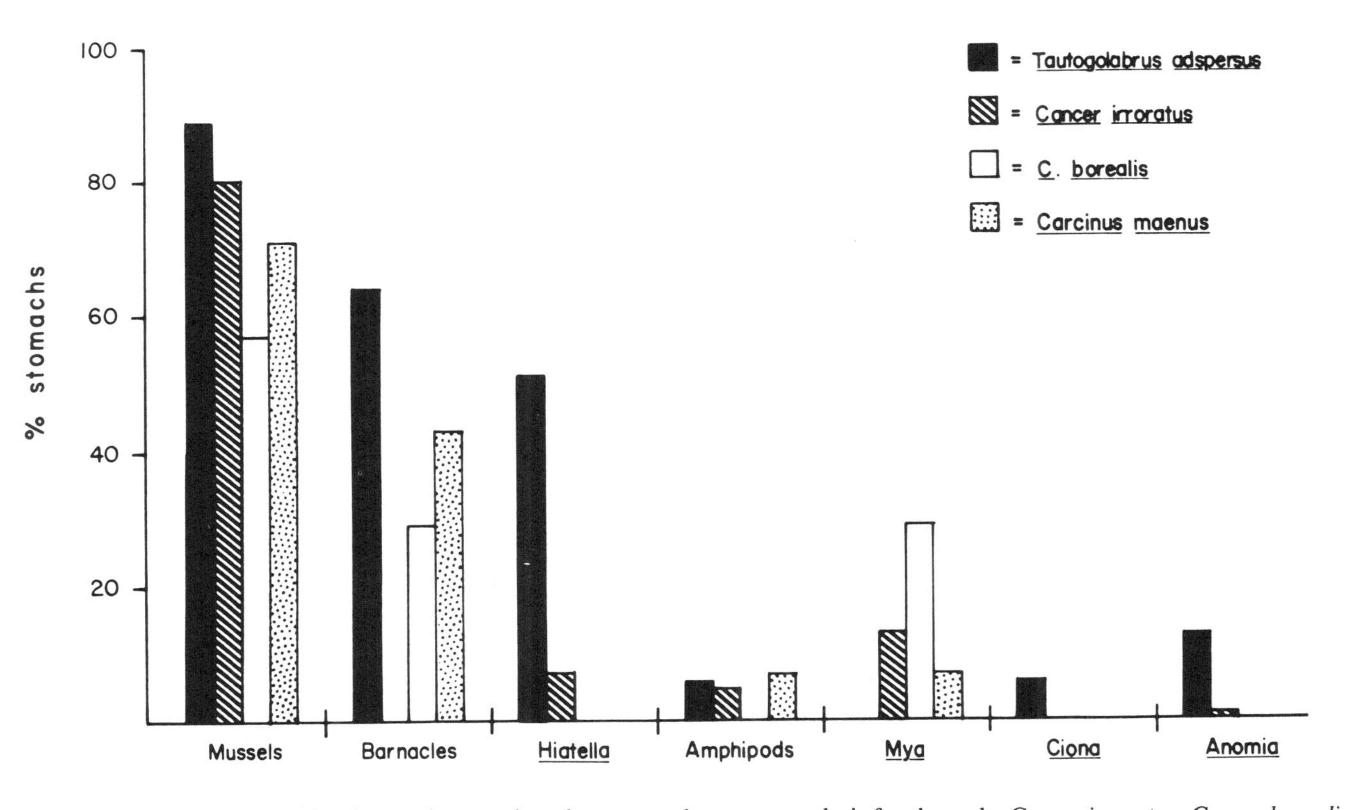

Figure 6. Comparison of feeding preferences based on stomach content analysis for the crabs *Cancer irroratus*, *Cancer borealis*, *Carcinus maenus* and the wrasse *Tautogolabrus adspersus*.

Another group of predators also was observed regularly on the panels and its feeding preferences documented. Nudibranchs were common as was the sea star *Henricia sanquinolenta.* These predators specialized primarily on sessile, colonial invertebrates such as sponges, ectoprocts and hydroids. *Henricia* fed primarily on *Haliclona* sp. and was the only sponge predator observed during the study. The dorid nudibranchs *Onchidoris muricata* and *Polycera dubia* fed on encrusting and upright ectoprocts, respectively. *Dendronotus frondosus, Facelina bostoniensis, Coryphella verrucosa* and *Catriona aurantia* all were observed feeding on *Tubularia* spp. *Aeolidia papillosa* was present during the winter as a predator on *Metridium senile.* The sea star *Henricia* was only found on the predator-access panels while the nudibranchs were present on all panel sets. A characteristic of nudibranchs, particularly the hydroid predators, was their continued residence on the panels after the obvious colonies of hydroids had been killed; presumably they were existing on newly settled polyps and, therefore, inhibiting reestablishment of hydroids after the initial heavy settlement.

Competitive Interactions

Analysis of direct encounters between the major sessile species with broad basal attachments indicated some clear patterns (Table VII, Figure 7). The majority of all encounters were standoffs in which the two organisms stopped growing at the point of contact. Encrusting ectoprocts were observed to erect walls against sponges and *Botryllus* and to be overgrown from the rear so that direction of encounter was important. *Anomia* species were effective at inhibiting overgrowth around most of the shell, but not from the direction of the hinge. *Anomia* appeared to be capable of inflicting physical damage on the growing edge

Table VII. Summary of Competitive Interactions for Each Species Involved

	Total Number Interactions	% Win	% Loss	% Standoff
Sponges	201	41	0	59
Botryllus	318	41	1	58
Ectoprocts	211	12	34	54
Metridium	13	8	0	92
Anomia	376	1	28	71
Balanus	104	0	59	41

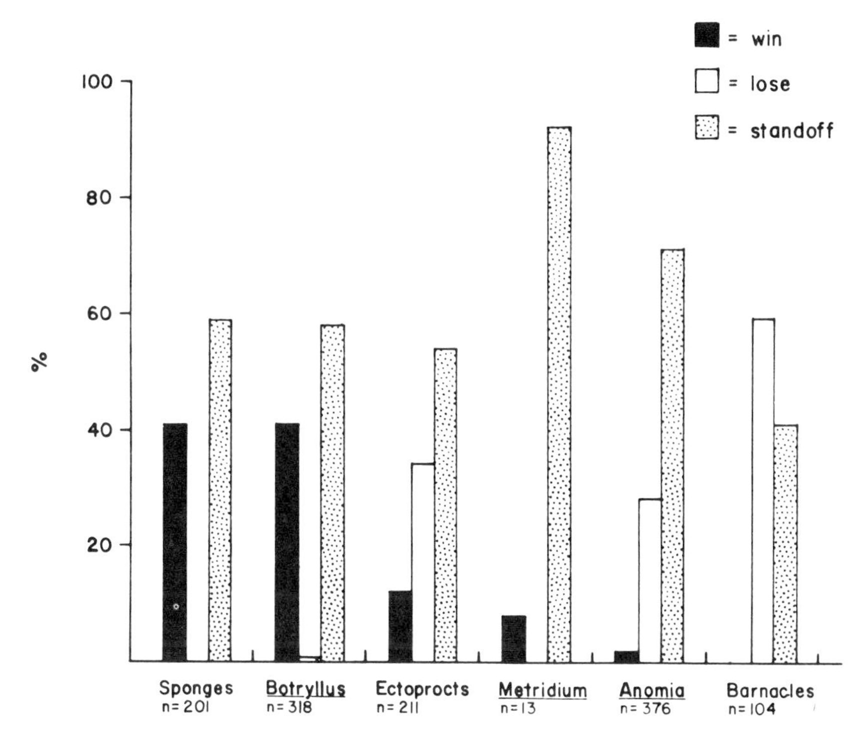

Figure 7. Results of observations of competitive encounters for those sessile species with encrusting growth forms or broad basal attachments.

of encrusting forms encroaching on the valves; this was most obvious with ectoprocts in which the abraded skeleton was conspicuous.

The only species that lost a significant number of encounters were encrusting ectoprocts, *Anomia* and *Balanus*. Sponges, *Botyllus* and *Metridium* lost few or no encounters and were winners in many encounters.

The patterns of who wins and loses in these competitive encounters suggest that surface characteristics may play a role in determining the outcome of an encounter. Table VIII and Figure 8 summarize the results of encounters between organisms with a soft, exposed tissue cover and those with hard outer surfaces. Soft organisms either win or stand off with species with hard coverings and mostly tie when encountering other species with soft surfaces. Animals with hard outer surfaces appear to stand off in most encounters, with only ectoprocts showing much ability to overgrow and then only against other forms with hard surfaces.

The one species that is an effective space competitor while having a

Table VIII. Summary of Competitive Interactions Between Organisms Related to Body Surface Characteristics (soft = sponges, *Botryllus, Metridium*; hard = encrusting ectoproct, *Anomia, Balanus*)

	Total Number Interactions	% Win	% Loss	% Standoff
Soft vs Hard	720	58	0	42
Soft vs Soft	172	2	2	96
Hard vs Hard	333	8	10	82

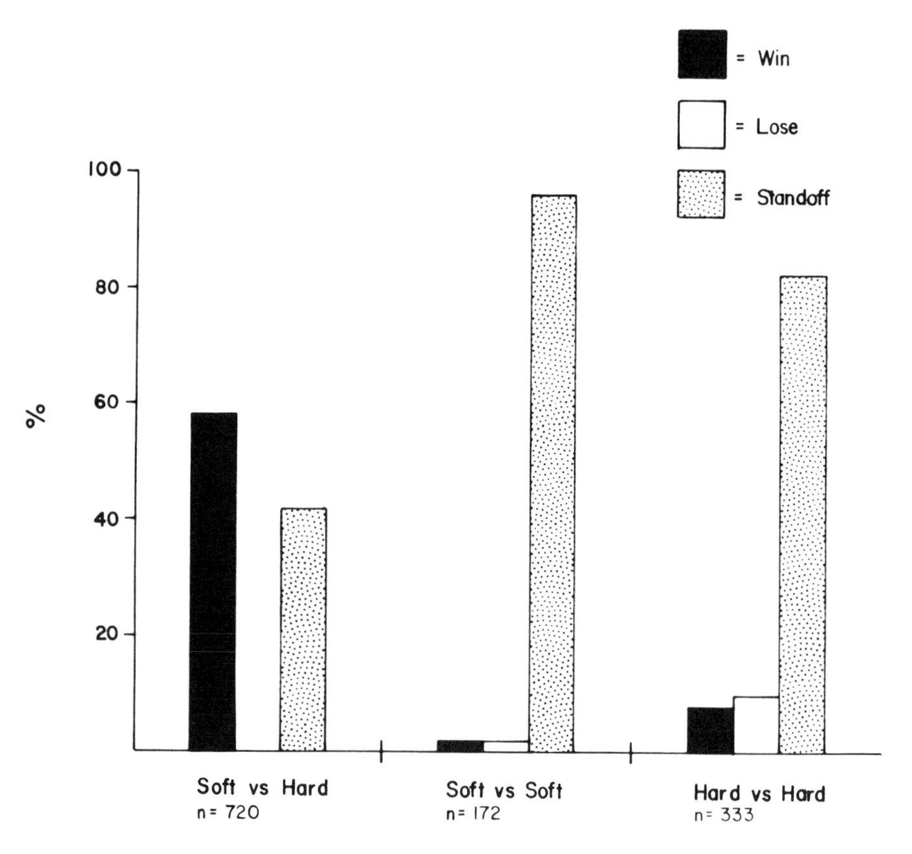

Figure 8. Summary of outcomes in encounters between sessile species based on their surface covering. Soft forms had a tissue layer or soft organic tunic as an outer covering while the hard forms had a hardened, non-living outer covering.

hard shell is the blue mussel *Mytilus edulis*. However, the mechanism of competition is distinct in that it requires clumping of mussels that expand through growth and continued aggregation. Mussels are raised off the substrate, being attached by byssal threads, so that overgrowth often occurs with minimal actual contact with encrusting forms. Sediment accumulation within the clumps appears to be the major means of mortality for animals overgrown by mussels.

The patterns of space occupation by the four major space-occupiers on the panels are summarized in Table IX and Figure 9. On first-year panels (Ia and II), *Botryllus* and encrusting ectoprocts are the most conspicuous encrusting forms, although sponges were more common on II, presumably due to the longer submersion time. Mussels on predator exclusion surfaces and sponges on lower horizontal and vertical surfaces became increasingly important in the second year (compare Ia and Ib). In fact, sponges seem to be more resistant to mussel overgrowth than any of the other species common on the panels.

DISCUSSION

The principal focus of this study was to test the effects of substrate angle and predation on fouling community succession. The results to date in this continuing research strongly indicate that substrate angle and predation influence the development of communities composed of sessile suspension-feeding invertebrates. The effects of these two factors can be observed to act separately as well as in an integrated manner to alter both species composition and structural characteristics of the communities present. The implications of the results described previously will be discussed below.

Substrate Angle

Species composition and relative abundance are clearly demonstrated to vary according to substrate angle (Table III and Figure 2). Upper horizontal surfaces were characterized by detritus accumulations that inhibited survival of species with encrusting growth forms, the exceptions being *Anomia* and *Crepidula* species on predator-access panels where crabs dispersed the silt enough to facilitate survival of these two molluscs. Both *Anomia* and *Crepidula* are found on silt-covered rocks below the panels, presumably indicating a high tolerance for sedimentation.

Table IX. Summary of Percent Cover of Forms with Encrusting, Colonial Growth for the Three Data Sets of Whole Panel Photographs

Encrusting Form	Access						Exclusion					
	UH		LH		V		UH		LH		V	
	$\bar{x}$	%	$\bar{x}$	%	$\bar{x}$	%	$\bar{x}$	%	$\bar{x}$	%	$\bar{x}$	%
Ia												
Sponges	0.0	0	5.1	2	3.4	1	0.5	0	5.9	2	2.1	1
Botryllus	1.5	1	38.5	14	87.5	32	0.4	0	67.1	25	84.6	31
Ectoprocts	0.0	0	12.6	5	12.3	5	0.0	0	9.9	4	9.2	3
$\Sigma\bar{X}$	1.5		56.2		103.2		0.9		82.9		95.9	
Total % Cover		1		21		38		0		30		35
Ib												
Sponges	6.6	2	68.8	25	17.1	6	0.3	0	49.1	18	20.6	8
Botryllus	0.4	0	21.5	8	13.9	5	0.0	0	13.5	5	9.2	3
Ectoprocts	0.4	0	2.6	1	1.0	0	0.0	0	0.4	0	0.0	0
$\Sigma\bar{X}$	7.4		92.9		32.0		0.3		63.0		29.8	
Total % Cover		3		34		12		0		23		11
II												
Sponges	0.1	0	7.1	3	7.0	3	0.1	0	29.6	11	31.7	12
Botryllus	2.4	1	41.1	15	68.6	25	0.1	0	48.6	18	76.0	28
Ectoprocts	0.9	0	11.7	4	11.5	4	0.1	0	5.4	2	7.9	3
$\Sigma\bar{X}$	3.4		59.9		87.1		0.3		83.6		115.6	
Total % Cover		1		22		32		0		31		42

[a] The mean number of points per plate is given for sponges, *Botryllus* and ectoprocts, as is the $\bar{x}$ percent cover per plate, and the total percent cover for all three groups by surface angle and location: Ia = November 1979 for first panel set; Ib = November 1980 for first set; II = November 1980 for second set.

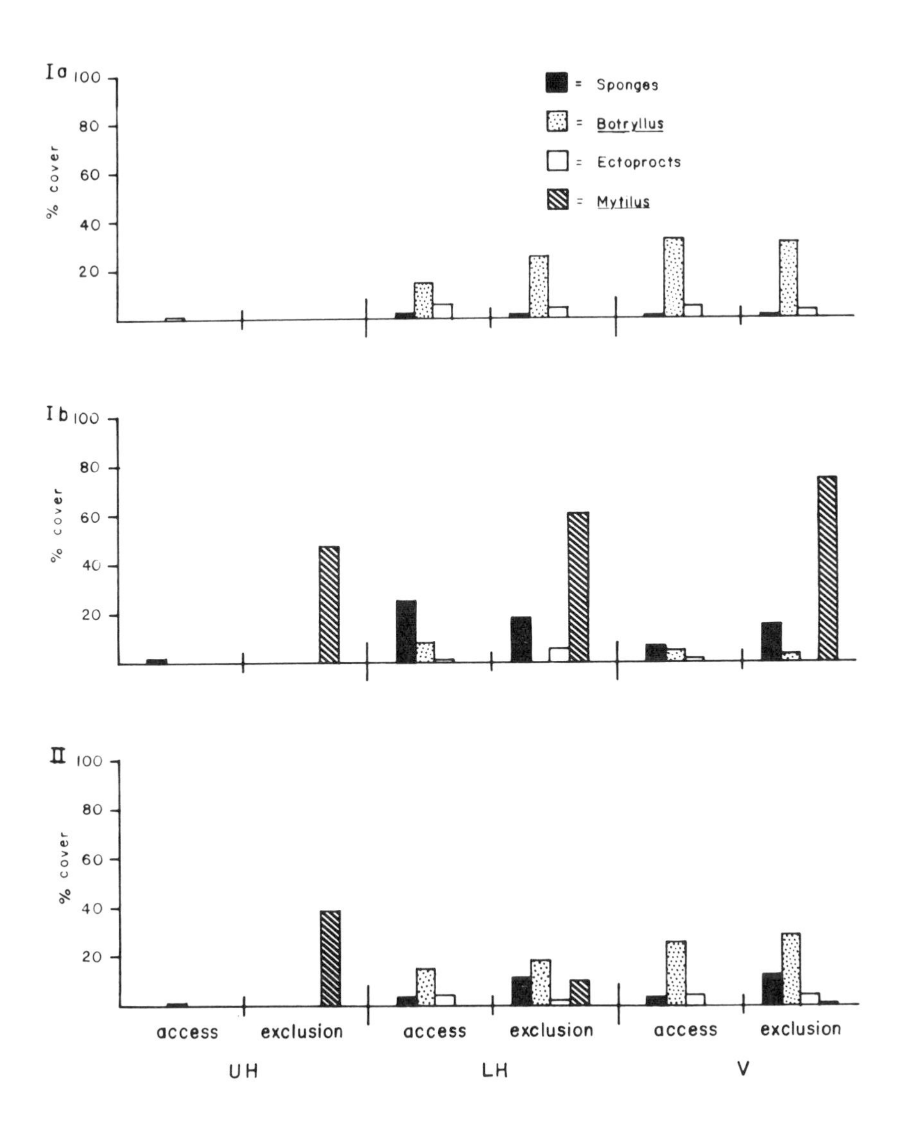

Figure 9. Summary of patterns of percent cover for the four most common space occupiers on the panels. The data is presented for each sampling period by predator access and substrate angle.

The other major space-occupiers on upper surfaces, thecate hydroids and *Mytilus edulis*, both have upright growth forms, which should make them resistant to siltation. *Mytilus* is primarily responsible for the majority of sediments on mussel-dominated surfaces through the production of feces and pseudofeces. Nothing but terebellid polychaetes have been observed living within the sediments on mussel-covered panels, indicating that one of the mechanisms by which mussels exclude potential competitors with encrusting growth forms is by smothering. This would be most effective on upper horizontal surfaces, and the absence of silt-tolerant species like *Anomia* and *Crepidula* from the upper surfaces dominated by *Mytilus* is evidence of this mechanism.

Communities on upper horizontal surfaces tend to be dominated by organisms with upright and mounding growth forms [34,42,46,47,49,50, 58–60]. This appears to be adaptive to resist the dual factors of disturbance by grazing predators and sediment accumulation, much of which is of biogenic origin. The exception to this pattern on upper surfaces are the crustose coralline algae, which are resistant to grazing, burial by sediments and overgrowth [58–61]. If encrusting growth forms are excluded from upper surfaces due to biogenically produced sediment deposition, then this phenomenon is similar to the concept of trophic amensalism proposed by Rhoads and Young [62] for soft sediment communities. The success of *Mytilus* on the upper surfaces indicates the selective advantage of an upright growth form on this substrate angle and the effectiveness of trophic amensalism as a mechanism of competitive exclusion in at least some hard substrate systems.

Lower horizontal and vertical surfaces of our panels have developed assemblages of sessile and sedentary suspension-feeding invertebrates, which are typically associated with fouling communities. The structure of the communities on these two substrate angles is quite distinct, with more surface area dominated by species with upright growth forms on lower horizontal surfaces while vertical surfaces tend to have a greater proportion of encrusting forms; the exception to this being the second-year (Ib) predator-exclusion vertical panels, where mussels dominate. The presence of more upright forms on the lower horizontal surfaces provides more structural complexity to the community and more potential refuges for associated fauna and the young stages of sessile species [40].

It is difficult to generalize about the patterns of upright versus encrusting growth forms by substrate angle since these two patterns can be found on both substrate angles in most regions of the world, and it is quite likely that they can be considered potential alternate stable states [13,15,31], depending on other abiotic and biotic factors present.

However, the results of this study do suggest that substrate angle should be considered in evaluating the results of panel studies. Sutherland [15–18,20] worked with lower horizontal surfaces in his panel studies at Beaufort, North Carolina while Karlson [8,9] was investigating vertical piling communities adjacent to the panel arrays. Some of the differences in patterns they observed may have been influenced by substrate angle.

Predation

There is a very clear predation effect on the panels, as illustrated by the differences between predator-access and predator-exclusion panels. *Mytilus edulis* dominated on the predator exclusion panels; only one small clump of *Mytilus* was found on one lower horizontal surface of one of the predator-access panels. This is not surprising when one considers the documented preference of the predators in the system for mussels. *Asterias* [36,63–65], *Cancer* [37,55,66–70], *Homarus* [54] and *Tautogolabrus* [71,72] all show a strong preference for mussels. These reports are corroborated by field observations and analyses of stomach contents of *Cancer* and *Carcinus* and *Tautogolabrus* (Table VI, Figure 6). As Peterson [39] suggested from his studies on fouling communities on pilings, crabs, in addition to fish and sea stars, can effectively exclude mussels from subtidal communities where otherwise they can be the dominant space competitors.

This is similar to the situation in the rocky intertidal where sea stars have played a similar role in reducing competition from mussels, thereby increasing diversity [44,46]. However, in fouling communities elimination of mussels only appears to alter which organisms will become the competitive dominants to those groups with colonial, encrusting growth strategies [6,57].

Three types of predation effects have been observed on our panels and a fourth has been described by Karlson [9] and Vance [73]. The exclusion of *Mytilus edulis* from the predator-access panels represents an example of the keystone predator phenomenon [43], but with the impact being the result of a guild of predators concentrating on a common resource. Peterson [39] showed a similar result on piling communities along the New Jersey coast, in which crabs were the major predators. In the subtidal, the diversity of predators is such that a single species is unlikely to act alone; it is more likely that several predators may have a synergistic impact resulting in the control of a competitive dominant. For example, fish [74] and urchins [75] are equally capable of producing halos around patch reefs, and it is most likely that both groups contribute to the maintenance of halos similar to the situation in terrestrial systems [76].

Predation on mussels appears to be size-selective, with the heaviest mortality taking place on the smaller-sized classes [45]. This is particularly so for the wrasses, which must use their pharyngeal plates to break the shells of their prey [71]. Evidence for this selection is the inability of the wrasses, which did forage on the panels away from the crib to control the mussels. Crabs are also size-limited in their ability to open mussels, although both *Cancer* species are capable of opening mussels with a shell length equal to, or greater than, their carapace width if the mussels are not attached and the crabs are hungry enough [55]. *Asterias* also tends to be a size-selective predator [36]. Small *Asterias* that had settled on the panels during the summer were feeding primarily on *Anomia* and small *Mytilus* that recruited during the summer.

As mentioned previously, most studies of *Mytilus edulis* have documented summer and early fall recruitment [77–79]. Lack of information on winter and early spring recruitment may be an artifact of seasonal, short-term panel studies or on a mild winter. An alternative hypothesis is that winter and spring settling is an adaptation to avoid the intensive larval filter [50] awaiting recruiting bivalves during the summer. By settling before most predators have become active and during the period when *Balanus balanoides* young are swamping the predators present, young mussels may have time to attain a moderate size refuge and, therefore, survive in otherwise unsuitable habitats. Winter recruitment is unlikely to occur in intertidal communities because of the harsh physical conditions.

An additional suggestion of size-selective predation is the presence of *Ciona* only on lower horizontal surfaces and primarily on the predator-exclusion panels. Sutherland [15] and Russ [40] have demonstrated that fish feed on small tunicates, and Russ has also shown that cover (artificial hydroids) resulted in increased survival of tunicates. Lower horizontal panels were characterized by abundant hydroid colonies and this upright cover presumably provides a favorable environment for *Ciona* survival by reducing the foraging efficiency of the wrasses, which do eat small *Ciona* (Table VI).

Substrate angle also appears to influence predator foraging efficiency. Crabs and large *Asterias* seldom were observed on lower horizontal surfaces, presumably because gravity was working against them. Wrasses also seem to be less inclined to forage on the ceilings of caves or the lower horizontal surfaces of the panels than on vertical or upper surfaces.

Cropping predators such as opisthobranchs and pycnogonids, which fed on hydroids and ectoprocts on our panels, are a conspicuous component of temperate fouling communities where they act as predators on barnacles, sponges, hydroids, anemones, ectoprocts and tunicates [80–83]. *Onchidoris muricata* has been observed feeding on encrusting ectoprocts, while *Polycera dubia* has been found on upright ectoprocts. Hydroids, however, appear to receive the most intensive predation pressure of nudibranchs. Most hydroids have an opportunistic strategy of heavy recruitment, fast growth and early initiation of reproduction. We believe the basis for this life history strategy is the result of strong selective pressure from predation rather than competition. Hydroids have specialized predators (nudibranchs and pycnogonids), which recruit onto established colonies and overwhelm their prey over time in the manner of a parasitoid or an insect herbivore. Most hydroid-eating nudibranchs and pycnogonids have fast growth rates, are small (no more than a few mm in body length) and feed by either cropping polyps one at a time or piercing the perisarc and sucking out the coenosarc. These feeding and life history strategies result in a race in which the prey grows and reproduces while predators increase in size and number. With increasing predation, the growth rate of the hydroid colony becomes negative and the colony is finally overwhelmed. For example, colonies of *Tubularia larynx* on our panels were simultaneously attacked by a pycnogonid species and the nudibranchs *Catriona aurantia, Coryphella verrucosa, Dendronotus frondosus* and *Facelina bostoniensis.* While most hydroid species have fewer nudibranch species that prey on them than *T. larynx*, the mechanism of accumulating predation pressure over time appears to be a consistent pattern. Turner [84] has found a small cuthonid nudibranch that presumably eats hydroids on wood panels in the deep sea.

Nudibranchs are typically small and cryptic in coloration and behavior so that they are inconspicuous until they have overwhelmed their prey and are out foraging, resulting in the sudden appearances described in the literature [4,80]. As with hydroids, colonial growth in general may be an effective defense against cropping predators, as partial predation does not kill the prey. Also, the encrusting growth form typical of many effective space competitors may serve to expose nudibranchs to fish. Harris [85] has experimentally demonstrated size-selective predation on the anemone-eating nudibranch *Aeolidia papillosa* by *Tautogolabrus.* Harris [85] has suggested that size-selective predation on cropping predators by visual predators may have suppressed the coevolution of these specialists, with those colonial forms adapted to tropical open fouling communities allowing the evolution of complex competitive mechanisms described recently [6,22,38,57]. Mechanisms that have selective value for competition as well as for predator defense are likely to be more successful than mechanisms having a single advantage.

Karlson [9] and Vance [73] have demonstrated that indiscriminate grazing by sea urchins can be a form of disturbance that increases diversity by opening space for recruitment in established communities. Mechanical damage caused by the foraging activities of crabs on the predator-access panels may have a similar effect by inhibiting the development of sponges on vertical surfaces.

Competitive Interactions

Competitive mechanisms such as those described by Jackson [6], Jackson and Buss [38] and Woodin and Jackson [57] were easily documented on the panels.

The angle of encounter was an important factor in the competitive interactions. *Anomia* was able to maintain itself from frontal assault by all of the encrusting, colonial forms. The ectoproct *Amphiblestrum flemingii* withstood frontal assault by sponges and *Botryllus* by erecting a wall, but it was readily overgrown from the rear. When *Amphiblestrum* encountered *Anomia* in a frontal assault there was physical damage obvious at the edge of the ectoproct colony adjacent to the *Anomia* valves.

Surface cover appears to be a factor that has not been dealt with adequately. The groups represented in this study generally fall into two categories: (1) those with a hard, nonliving outer surfaces, and (2) those with a tissue layer exposed. Sponges, anthozoans and some tunicates have tissue layers exposed, while hydroids, ectoprocts, barnacles, bivalves and *Spirorbis* all have some form of hard outer covering, presumably as a defense against predation. As is demonstrated in Table VIII, animals with a hard outer surface can be overgrown, while animals with a soft outer layer cannot. Species like *Anomia* or *Balanus* may survive with the shell overgrown as long as they can still feed but ultimately are likely to succumb to smothering or be outcompeted for food. Woodin and Jackson [57] suggested that resource partitioning of food may take place between suspension feeding members of fouling communities. As an example, ectoprocts generally feed on flagellates [86], while sponges eat primarily submicroscopic particles [87]. The problem with this suggestion is that Reiswig [87] found that sponges filter out everything from the water, including submicroscopic materials, leaving nothing for less-efficient suspension feeders.

The advantage of a soft surface was further demonstrated by the fact that hummocking of mussels occurred in the area of many panels where sponge colonies were already established, reducing the substrate available to attach byssal threads. In fact, sponges appeared to do best where interactions with mussels reduce competition from other encrusting forms (Table IX).

One other factor that appears to be important in the competitive interactions is longevity. Encounters between sponges and *Botryllus* invariably resulted in a standoff, but *Botryllus* colonies live only a few months, undergo senescence and drop out of the system. The sponges therefore can afford to wait for free space to open up, a strategy similar to that documented for *Hydractinia echinata* [9]. Ectoprocts erect walls against overgrowth by *Botryllus* and sponges, but this only buys time as the ectoprocts cannot overgrow a sponge or tunicate. While a few species of ectoprocts have life spans longer than one year [13,16–18], most species are short-lived and are merely resisting overgrowth to maximize reproduction. Woodin and Jackson [57] reported that the majority of the oldest space under coral plates where competitive networks between ectoprocts have been documented are actually dominated by sponges.

Sponges appear to be ideally suited to be the ultimate competitive dominants on vertical and undercut surfaces. They are extremely efficient suspension feeders, are long-lived, grow to relatively large sizes, have plasticity of growth form and high powers of regeneration from cropping predators, and have a soft outer surface that is resistant to overgrowth by anything but other sponges. In addition, they are being shown to produce an impressive array of allelochemicals [38,88,89].

Therefore, it is not surprising that sponges are conspicuous members of most fouling communities and ultimately tend to dominate many such systems given time and lack of disturbance. Harris has observed subtidal communities in a number of geographic locations, and a review of habitats confirms that sponges are major space-occupiers of fouling communities in tropical [11,19,22,57,90,91], temperate [9,49] and subarctic environments [26].

Predators may have facilitated the competitive sequence seen on the panels by selective predation on hydroids and ectoprocts. *Onchidoris muricata* preyed on *Amphiblestrum* and several nudibranchs were predators on *Tubularia* spp., which formed dense aggregations during the first summer each panel array was established (Figure 2). Standing [27] demonstrated that hydroids could effectively hold space for long periods of time, and Karlson [9] showed that the hydroid *Hydractina echinata* is a superior long-term space-occupier that expands as disturbance produces free space at the periphery of the colony. Russ [40] has shown that artificial hydroids provide a refuge for young tunicates and ectoprocts from fish predation. It is possible that actively growing colonies of hydroids are resistant to settlement by potential space competitors, the dense aggregation of feeding polyps forming an effective larval filter. However, nudibranch predation on polyps would open the colony to settlement of recruiting larvae and the hydroid colony then becomes a refuge from size-selective visual predators, thus facilitating succession in these communities.

Substrate Composition and Texture

The sloughing off of *Mytilus* from all of the vertical predator-free panels set up in February 1980 may have been due to the smoothness of the plexiglass used. Certainly surface texture can influence settlement of some species, especially in short-term studies [92–95]. However, observations of natural substrates, either newly implanted or kept free of grazers [77], still are not comparable in community organization to the surrounding communities after two and one years, respectively. For long-term studies, therefore, substrate type may be less important than factors such as substrate size, angle and location.

Synergism

A number of factors have been identified as influencing the development of fouling communities. A goal of ecology is to attempt to identify general patterns that will provide a unified theory. At the same time, the complexity of communities requires a reductionist approach. As a result there is a tendency to focus on obvious factors such as competition within a taxonomic group to the partial or complete exclusion of other groups and factors [96]. As an example, colonial growth has definite value as a competitive mechanism [77], but is this any more important than as a defense against cropping predators? It is also possible that having many small feeding and reproductive units to maximize feeding efficiency on small particles reduces the energy expenditure of maintaining a large mass so that a majority of energy acquired can be used for reproduction [97].

In summary, the fouling community studies described and discussed here provide a verification of the model of community succession proposed by Connell [31]. Given time, fouling communities do tend to be dominated by a few effective competitors. A variety of factors, including patch size, substrate angle, predation and characteristics of the competitors all influence the successional sequence and determine who will ultimately dominate that patch.

ACKNOWLEDGMENTS

The research described above was supported by grants from the Office of Naval Research and Sea Grant. Help in preparation of the manuscript by Wendy Lull, Gary Gaulin and Theodore Donn is gratefully acknowledged.

REFERENCES

1. Boyd, M. J. "Fouling Community Structure and Development in Bodega Harbor, California," Ph.D. Thesis, University of California, Davis, CA (1972).
2. Coe, W. R., and W. E. Allen. "Growth of Sedentary Marine Organisms on Experimental Blocks and Plates for Nine Successive Years at the Pier of the Scripps Institution of Oceanography," *Bull. Scripps Inst. Oceanog., Tech. Ser.* 4: 101–136 (1937).
3. Cory, R. L. "Epifauna of the Patuxent River Estuary, Maryland, for 1963 and 1964," *Chesapeake Sci.* 8: 71–89 (1967).
4. Fager, E. W. "Pattern in the Development of a Marine Community," *Limnol. Oceanog.* 16: 241–253 (1971).
5. Fuller, J. L. "Season of Attachment and Growth of Sedentary Marine Organisms at Lamoine, Maine," *Ecol.* 27: 150–158 (1946).
6. Jackson. J. B. C. "Competition on Marine Hard Substrata: the Adaptive Significance of Solitary and Colonial Strategies," *Am. Nat.* 111: 743–767 (1977).
7. Buss, L. W., and J. B. C. Jackson, "Competitive Networks: Nontransitive Competitive Relationships in Cryptic Coral Reef Environments," *Am. Nat.* 113:223–234 (1979).
8. Karlson, R. H. "The Effects of Predation by the Sea Urchin, *Arbacia punctulata,* on a Marine Epibenthic Community," Ph.D. Thesis, Duke University, Durham, NC (1975).
9. Karlson, R. H. "Predation and Space Utilization Patterns in a Marine Epifaunal Community," *J. Exp. Mar. Biol. Ecol.* 31: 225–239 (1978).
10. Keough, M. J., and A. J. Butler. "The Role of Asteroid Predators in the Organization of a Sessile Community on Pier Pilings," *Mar. Biol.* 51: 167–177 (1979).
11. McDougall, K. D. "Sessile Marine Invertebrates of Beaufort, North Carolina," *Ecol. Monog.* 13: 321–374 (1943).
12. Nair, N. B. Ecology of Marine Fouling and Boring Organisms of Western Norway," *Sarsia* 8: 1–88 (1962).
13. Osman, R. W. "The Establishment and Development of a Marine Epifaunal Community," *Ecol. Monog.* 47: 37–63 (1977).
14. Scheer, B. T. "The Development of Marine Fouling Communities," *Biol. Bull.* 89: 103–121 (1945).
15. Sutherland, J. P. "Multiple Stable Points in Natural Communities," *Am. Nat.* 108: 859–873 (1974).
16. Sutherland, J. P. "Effect of *Schizoporella* Removal on the Fouling Community at Beaufort, N.C." in *Ecology of Marine Benthos.* B. C. Coull, Ed. Belle W. Baruch Institute for Marine Biology and Coastal Research Symposium (Columbia, SC: University of South Carolina Press, 1975), p. 155.
17. Sutherland, J. P. "Life Histories and Dynamics of Fouling Communities," in *Proc. U.S.-U.S.S.R. Workshop on the Ecology of Fouling Communities,* Office of Naval Research (1976), pp. 137-153.
18. Sutherland, J. P. "Functional Roles of *Schizoporella* and *Styela* in the Fouling Community at Beaufort, North Carolina," *Ecology* 59: 257-264 (1978).

19. Sutherland, J. P. "Dynamics of the Epibenthic Community on Roots of the Mangrove *Rhizophora mangle*, at Bahia de Buche, Venezuela," *Mar. Biol.* 58: 75–84 (1980).
20. Sutherland, J. P., and R. H. Karlson. "Development and Stability of the Fouling Community at Beaufort, North Carolina," *Ecol. Monog.* 47: 425–446 (1977).
21. Weiss, C. M. "The Seasonal Occurrence of Sedentary Marine Organisms in Biscayne Bay, Florida," *Ecology* 29: 153–172 (1948).
22. Wells, H. W., M. J. Wells and I. E. Gray. "Ecology of Sponges in Hatteras Harbor, North Carolina," *Ecology* 45: 752–767 (1964).
23. Woods Hole Oceanographic Institution. "Marine Fouling and its Prevention," U.S. Naval Institute, Annapolis, MD (1952).
24. Graham, H. W., and H. Gay. "Season of Attachment and Growth of Sedentary Marine Organisms at Oakland, California," *Ecology* 26: 375–386 (1945).
25. Normandeau Associates, Inc. "Piscataqua River Ecological Studies, 1975 Monitoring Studies," Report No. 6 for Public Service Company of New Hampshire (1976).
26. Pelletier, P. "Zonation and Biomass of Piling Epifauna at the Mouth of a Boreal Estuary in Winter," Zoology 803 report, University of New Hampshire, Zoology Department (1972).
27. Standing, J. D. "Fouling Community Structure: Effects of the Hydroid, *Obelia dichotoma,* on Larval Recruitment," in *Coelenterate Ecology and Behavior,* G. O. Mackie, Ed. (New York: Plenum Publishing Corp., 1976), p. 155.
28. Hoffman, R. J. "Genetics and Asexual Reproduction of the Sea Anemone *Metridium senile,*" *Biol. Bull.* 151: 478–488 (1976).
29. Purcell, J. E. "Aggressive Function and Induced Development of Catch Tentacles in the Sea Anemone *Metridium senile* (Coelenterata, Actiniaria)," *Biol. Bull.* 153: 355–368 (1977).
30. Harris, L. G. "Comparative Ecological Studies on the Nudibranch *Aeolidia papillosa* and its Anemone Prey *Metridium senile* along the Atlantic and the Pacific coasts of the United States," *J. Moll. Studies* 42: 301 (1976).
31. Connell, J. H. "Some Mechanisms Producing Structure in Natural Communities: A Model and Evidence from Field Experiments," in *Ecology and Evolution of Communities,* M. L. Cody and J. Diamond, Eds. (Cambridge, MA: Belknap Press, 1975).
32. Connell, J. H. "Diversity in Tropical Rain Forests and Coral Reefs," *Science* 199: 1302–1310 (1978).
33. Janzen, D. H. "Herbivores and the Number of Tree Species in Tropical Forests," *Am. Nat.* 104: 501–528 (1970).
34. Jones, O. A., and R. Endean, Eds. *Biology and Geology of Coral Reefs,* Vol. I, Geology 1 (New York: Academic Press, Inc., 1973).
35. Sanders, H. L. "Marine Benthic Diversity: a Comparative Study," *Am. Nat.* 102: 243–282 (1968).
36. Hulbert, A. W. "The Functional Role of *Asterias vulgaris* Verrill (1866) in Three Subtidal Communities," Ph.D. Thesis, University of New Hampshire, Durham, NH (1980).

37. Richardson, E., J. Perez, A. Hulbert, L. Black and K. Tacy. "Role of Predation in Subtidal Community Zonation," Sea Grant Program Report, University of New Hampshire, Durham, NH (1977).
38. Jackson, J. B. C., and L. Buss. Allelopathy and Spatial Competition Among Coral Reef Invertebrates," *Proc. Nat. Acad. Sci. U.S.* 72: 5160–5163 (1975).
39. Peterson, C. H. "The Importance of Predation and Competition in Organizing the Intertidal Epifaunal Communities of Barnegat Inlet, New Jersey," *Oecologia* (Berl.) 39: 1–24 (1979).
40. Russ, G. "Effects of Predation by Fishes, Competition, and Structural Complexity of the Substratum on the Establishment of a Marine Epifaunal Community," *J. Exp. Mar. Biol. Ecol.* 42: 55–69 (1980).
41. Annala, J. "Foraging Strategies and Predation Effects of *Asterias rubens* and *Nucella lapillus,*" Ph.D. Thesis, University of New Hampshire, Durham, NH (1974).
42. Menge, B. A. "Organization of the New England Rocky Intertidal Community: Role of Predation, Competition, and Environmental Heterogeneity," *Ecol. Monog.* 46: 355–393 (1976).
43. Paine, R. T. "Food Web Complexity and Species Diversity," *Am. Nat.* 100: 65–75 (1966).
44. Paine, R. T. "Intertidal Community Structure. Experimental Studies on the Relationship Between a Dominant Competitor and its Principal Predator," *Oecologia* (Berl.) 15: 92–120 (1974).
45. Paine, R. T. "Size-limited Predation: an Observational and Experimental Approach with the *Mytilus-Pisaster* Interaction," *Ecology* 57: 858–873 (1976).
46. Dayton, P. K. "Competition, Disturbance, and Community Organization: the Provision and Subsequent Utilization of Space in a Rocky Intertidal Community. *Ecol. Monog.* 41: 351–389 (1971).
47. Golikov, A. N., and O. A. Scarlato. "Comparative Characteristics of some Ecosystems of the Upper Regions of the Shelf in Tropical, Temperate and Arctic Waters," *Helgoländer Wiss. Meeresunters* 24: 219–234 (1973).
48. Harris, L. G., A. Hulbert, J. Witman, K. McCarthy and K. Pecci. "A Comparison by Depth and Substrate Angle of Subtidal Benthic Communities at Pigeon Hill in the Gulf of Maine," *Am. Zool.* 19(3): 1010 (1979).
49. Witman, J., A. Hulbert, L. Harris, K. Pecci, K. McCarthy and R. Cooper. "Community Structure of the Macrobenthos of Pigeon Hill in the Gulf of Maine," UNH-NMFS Ocean Pulse Technical Report (1980).
50. Dayton, P. K., G. A. Robilliard, R. T. Paine, and L. B. Dayton. "Biological Accommodation in the Benthic Community at McMurdo Sound, Antarctica," *Ecol. Monog.* 44: 105–128 (1974).
51. Norall, T. L., and A. C. Mathieson. "Nutrient and Hydrographic Data for the Great Bay Estuarine System and the Adjacent Open Coast of New Hampshire-Maine," Jackson Estuarine Laboratory Contribution (1976).
52. Glibert, P. M. "Nutrient Distribution Within the Tidal Rivers of the Great Bay Estuary System," Jackson Estuarine Laboratory Contribution (1976).
53. Gosner, K. L. *Guide to Identification of Marine and Estuarine Invertebrates.* (New York: Wiley-Interscience, 1971).

54. Squires, H. J. "Lobster *(Homarus americanus)* Fishery and Ecology in Port Au Port Bay, Newfoundland, 1960–65," *Proc. Nat. Shellfish Assoc.* 60: 22–39 (1970).
55. Harris, L. G., R. Langan, S. Massicotte and J. Perez. "Population Structure and Feeding Biology of two *Cancer* species Along an Exposed Subtidal Transect," Unpublished results (1980).
56. Green, J. M., and M. Farwell, "Winter Habits of the Cunner *Tautogolabrus adspersus* (Walbaum, 1792), in Newfoundland," *Can. J. Zool.* 49: 1497–1499 (1971).
57. Woodin, S. A., and J. B. C. Jackson, "Interphyletic Competition Among Marine Benthos," *Am. Zool.* 19: 1029–1043 (1979).
58. Lubchenco, J., and J. Cubit. "Heteromorphic Life Histories of Certain Marine Algae as Adaptations to Variations in Herbivory," *Ecology* 61: 676–687 (1980).
59. Littler, M. M., and D. S. Littler. "The Evolution of Thallus Form and Survival Strategies in Benthic Marine Macroalgae: Field and Laboratory Tests of a Functional Form Model," *Am. Nat.* 116: 25–44 (1980).
60. Harris, L. G., J. Derick, T. E. Donn, W. D. Lord and J. D. Witman. "Competitive Webs on Upper Horizontal Substrates: The Role of Grazers," Unpublished results (1979).
61. Steneck, R. S. "Factors Influencing the Distribution of Crustose Coralline Algae (Rhodophyta, Corallinaceae) in the Damariscotta River Estuary, Maine," MS Thesis, University of Maine, Orono, ME (1978).
62. Rhoads, D. C., and D. K. Young, "The Influence of Deposit-Feeding Organisms on Sediment Stability and Community Trophic Structure," *J. Mar. Res.* 28: 150–178 (1970).
63. Menge, B. A. "Coexistence Between the Seastars *Asterias vulgaris* and *A. forbesi* in a Heterogeneous Environment: A Non-Equilibrium Explanation," *Oecologia* 41: 245–272 (1979).
64. Hancock, D. A. "The Feeding Behavior of Starfish on Essex Oyster Beds," *J. Mar. Biol. Assoc. U.K.* 34: 313–331 (1955).
65. Hancock, D. A. "Notes on Starfish on an Essex Oyster Bed," *J. Mar. Biol. Assoc. U.K.* 37: 565–589 (1958).
66. Ebling, F. J., J. A. Kitching, L. Muntz and C. M. Taylor. "The Ecology of Lough Ine. XIII. Experimental Observations of the Destruction of *Mytilus edulis* and *Nucella lapillus* by Crabs," *J. Animal Ecol.* 33: 73–82 (1964).
67. Kitching, J. A., J. F. Sloane, and F. J. Ebling. "The Ecology of Lough Ine. VIII. Mussels and their Predators," *J. Animal Ecol.* 28: 331–341 (1959).
68. Muntz, L., F. J. Ebling, and J. A. Kitching. "The Ecology of Lough Ine. XIV. Predatory Activities of Large Crabs," *J. Animal Ecol.* 34: 315–329 (1965).
69. Seed, R. "The Ecology of *Mytilus edulis* L. (Lamellibranchiata) on Exposed Rocky Shores. II. Growth and Mortality," *Oecologia* 3: 317–350 (1969).
70. Walne, P. R., and G. J. Dean. "Experiments on Predation by the Shore Crab, *Carcinus maenus* (L.), on *Mytilus* and *Mercenaria*," *J. Conseil* 34: 190–199 (1972).
71. Chao, L. N. "Digestive System and Feeding Habits of the Cunner,

Tautogolabrus adspersus, a Stomachless Fish," *Fish. Bull.* 71: 565-586 (1973).

72. Shumway, S. E., and R. R. Stickney. "Notes on the Biology and Food Habits of the Cunner," *New York Fish Game J.* 22: 71-79 (1975).
73. Vance, R. R. "Effects of Grazing by the Sea Urchin, *Centrostephanus coronatus,* on Prey Community Composition," *Ecology* 60: 537-546 (1979).
74. Randall, J. E. "Grazing Effect on Sea Grasses by Herbivorous Reef Fishes in the West Indies," *Ecology* 46: 255-260 (1965).
75. Ogden, J. C., R. A. Brown, and N. Salesky. "Grazing by the Echinoid *Diadema antillarum* Philippi: Formation of Halos Around West Indian Patch Reefs," *Science* 182: 715-716 (1973).
76. Bartholomew, B. "Bare Zone Between California Shrub and Grassland Communities: The Role of Animals," *Science* 170: 1210-1212 (1970).
77. Harris, L. G., W. Lull, A. Bloomrosen and G. Gaulin. "The Effects of Urchin Removal in an Urchin Barrens Community," Unpublished results (1981).
78. Seed, R. "The Ecology of *Mytilus edulis* L. (Lamellibranchiata) on Exposed Rocky Shores." I. Breeding and Settlement," *Oecologia* 3: 277-316 (1969).
79. Seed, R., and R. A. Brown. "A Comparison of the Reproductive Cycles of *Modiolus modiolus* (L.), *Cerastoderma* (= *Cardium*) *edule* (L.), and *Mytilus edulis* L. in Strengford Lough, Northern Ireland," *Oecologia* 30: 173-188 (1977).
80. Harris, L. G. "Nudibranch Associations," in *Current Topics in Comparative Pathobiology,* T. C. Cheng, Ed. (New York: Academic Press, Inc., 1973). pp. 213-215.
81. Clark, K. B. "Nudibranch Life Cycles in the Northwest Atlantic and Their Relationship to the Ecology of Fouling Communities," *Helgolaňder Wiss. Meeresunters.* 27: 28-69 (1975).
82. Bloom, S. A. "Morphological Correlations Between Dorid Nudibranch Predators and Sponge Prey," *Veliger* 18: 289-301 (1976).
83. McDonald, G. R., and J. W. Nybakken. "Additional Notes on the Food of Some California Nudibranchs with a Summary of Known Food Habits of California Species," *Veliger* 21: 110-119 (1978).
84. Turner, R. Personal communication (1977).
85. Harris, L. G. "Size Selective Predation in a Sea Anemone, Nudibranch and Fish Food Chain," Unpublished results (1981).
86. Winston, J. E. "Feeding" in *The Biology of Bryozoans.* R. M. Woollacott and R. L. Zimmer, Eds., (New York: Academic Press, Inc., 1977), pp. 233-271.
87. Reiswig, H. M. "Particle Feeding in Natural Populations of Three Marine Demospongiae," *Biol. Bull.* 141: 568-591 (1971).
88. Green, G. "Ecology of Toxicity in Marine Sponges. *Mar. Biol.* 40: 207-215 (1977).
89. Randall, J. E., and W. D. Hartman. "Sponge-Feeding Fishes of the West Indies," *Mar. Biol.* 1: 216-225 (1968).
90. Rützler, K. "Spatial Competition among Porifera: Solution by Epizoism," *Oecologia* 5: 85-95 (1970).
91. Sará, M. "Competition and Cooperation in Sponge Populations," *Symp. Zool. Soc. Lond.* 25: 273-284 (1970).

92. Tippett, R. "Artificial Surfaces as a Method of Studying Populations of Benthic Micro-algae in Fresh Water," *Br. Phycol. J.* 5: 187–199 (1970).
93. Knight-Jones, E. W. "Gregariousness and Some Other Aspects of the Settling Behavior of *Spirorbis*," *J. Mar. Biol. Assoc. U.K.* 30: 201–222 (1951).
94. Knight-Jones, E. W. "Laboratory Experiments on Gregariousness During Settling in *Balanus balanoides* and Other Barnacles," *J. Exp. Biol.* 30: 584–598 (1953).
95. Stebbing, A. R. D. "Preferential Settlement of a Bryozoan and Serpulid Larvae on the Younger Parts of *Laminaria* Fronds," *J. Mar. Biol. Assoc. U.K.* 52: 765–882 (1972).
96. Dayton, P. K. "Ecology: a Science and a Religion" in *Ecological Processes in Coastal and Marine Systems,* R. J. Livingstone, Ed. (New York: Plenum Publishing Co., 1979), pp. 3–18.
97. Sebens, K. P. "The Energetics of Asexual Reproduction and Colony Formation in Benthic Marine Invertebrates," *Am. Zool.* 19: 683–697 (1979).

CHAPTER 6

THE USE OF ARTIFICIAL SUBSTRATES IN THE STUDY OF FRESHWATER BENTHIC MACROINVERTEBRATES

David M. Rosenberg

Department of Fisheries and Oceans
Freshwater Institute
Winnipeg, Manitoba R3T 2N6

Vincent H. Resh

Division of Entomology and Parasitology
University of California
Berkeley, California 94720

Claims in the literature of advantages and disadvantages of using artificial substrates to sample freshwater benthic macroinvertebrates were examined critically. Several advantages were found to be correct as claimed: Artificial substrates (1) are useful for collecting data from habitats difficult to sample by other methods; (2) reduce variability of operator efficiency in taking samples, thus helping to standardize a sampling program; (3) are relatively inexpensive and simple to construct, aspects that make them convenient to use; and (4) permit nondestructive sampling of an environment.

Several advantages were correct as claimed but required qualifications: (1) standardization of microhabitat by using artificial substrates depends on the type of sampler used; (2) artificial substrates increase sampling precision but may decrease sampling accuracy due to loss of information on spatial distribution of macroinvertebrates; (3) nonbiologists can participate effectively in sampling programs using artificial substrates, but potential problems must be considered; (4) certain aspects of the convenience of use of artificial substrates depend on the type of sampler used; and (5) artificial substrates yield more flexibility in sampling programs but this depends on the definition of "flexibility." The claim that artificial substrates help standardize a sampling program is only correct if artificial substrates are used in similar macrohabitats.

Two disadvantages were correct as claimed: (1) colonization of artificial substrates by macroinvertebrates is subject to seasonal variation (a problem of all sampling methods); and (2) relatively long exposure times are required to collect a sample. Little can be done to rectify these disadvantages.

Other disadvantages claimed are also correct: (1) times for populations to reach equilibrium levels in the colonization of artificial substrates are generally unknown; (2) artificial substrates are selective for organisms that colonize them; (3) macroinvertebrates are lost when artificial substrates are retrieved; and (4) artificial substances are subject to loss by vandalism, spate, drought or burial. However, many of these can be minimized with varying degrees of effort.

The remaining disadvantages claimed were correct but depended on the type of artificial substrate and how it was handled: (1) artificial substrates are not representative of the natural substrate or the effects of pollution on that substrate; and (2) artificial substrates are inconvenient to use and logistically awkward. These disadvantages also can be rectified.

There is a critical need to elucidate general principles governing the dynamics of macroinvertebrate colonization of artificial substrates. To this end we recommend that detailed studies of the operation of artificial substrates such as sequences of succession leading to colonization by macroinvertebrates, species and life stages of macroinvertebrates involved, biotic and abiotic factors affecting colonization, etc., should be encouraged, and that a research program be initiated to compare the operation of artificial substrates in various geographic locations. Until a better understanding of the operation of artificial substrates is achieved, pilot studies should be undertaken to calibrate use of artificial substrates at individual study sites.

INTRODUCTION

The use of artificial substrates for sampling macroinvertebrates originated from the realization that all aquatic habitats could not be sampled effectively using conventional devices such as grabs or nets (Beak et al. 1973). More recently, emphasis has been placed on using artificial substrates to reduce the variability associated with conventional sampling devices (Hellawell 1978).

The earliest uses of artificial substrates anticipated most contemporary uses. For example, Moon (1935a) surveyed the rocky littoral zone of Lake Windermere in England (a habitat difficult to sample quantitatively by direct means), with trays of rock or other types of substratum. Similarly, Wanson and Henrard (1945, in Disney 1972) used artificial substrates to standardize sampling of the blackfly larva, *Simulium,* which occurred on difficult-to-sample substrates.

Shelford and Eddy (1929) suggested that 1-m² concrete plaques or rocks be placed in swiftly flowing streams for the experimental study of the development and succession of stream invertebrates, and Moon (1935b, 1940) subsequently used the artificial substrates that he developed for Lake Windermere to study movements and colonization of new habitats by freshwater macroinvertebrates.

These studies anticipated experimental investigations of aspects of benthic macroinvertebrate ecology such as movement, colonization dynamics, feeding, microdistribution and substrate preference, topics of intense study today that have benefitted from the use of artificial substrates. Wene and Wickliff's (1940, p. 131) use of wire baskets filled with stream rubble "for studying the effects of contrasting types of stream bottom on the insect fauna" possibly anticipated the contemporary use of artificial substrates for survey activities such as water quality monitoring.

The use of artificial substrates for sampling macroinvertebrates has increased steadily over the past several years (Figure 1), and Hellawell (1978, p. 35), predicted that artificial substrate "methods are likely to become increasingly important in the routine surveillance of rivers." Recent reviews of artificial substrates have been mainly descriptive (e.g., Beak et al. 1973; Slack et al. 1973; Weber 1973; Lapchin 1977; Cover and Harrel 1978; Khalaf and Tachet 1978). However, that of Hellawell (1978) reassessed some published data. In this review we attempt to give a detailed and critical evaluation of the performance of this increasingly popular approach to sampling freshwater macroinvertebrates.

Objective

Our objectives are to summarize and evaluate the advantages and disadvantages, identified in the literature, of using artificial substrates for sampling freshwater macroinvertebrates; to discuss ways to improve sampling efficiency of artificial substrates; and to identify future needs

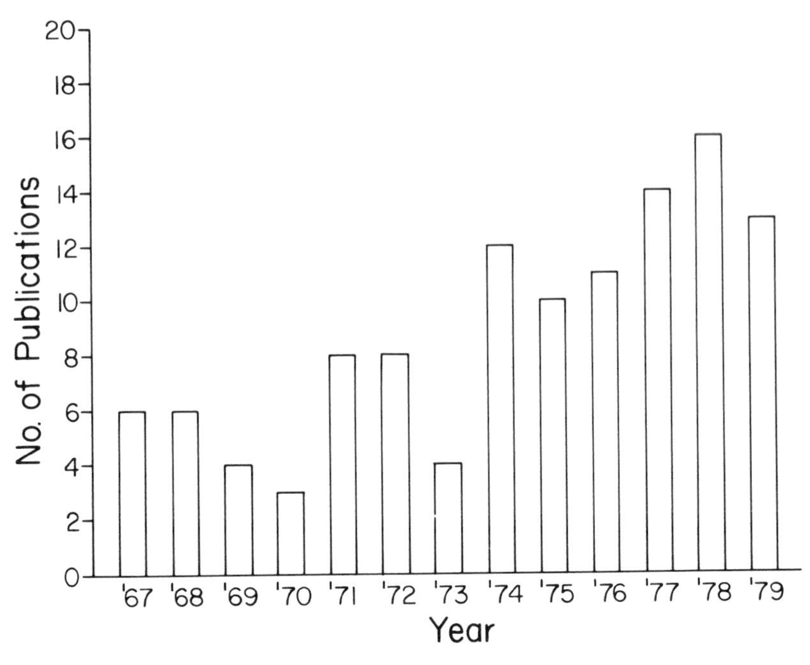

Figure 1. Numbers of publications devoted to artificial substrate samplers for benthic macroinvertebrates during the period 1967–1979 (Rosenberg 1978 and unpublished data).

in the use of artificial substrates. Decisions on their use will depend on study objectives, but we hope to present sufficient background information to enable someone designing a benthic macroinvertebrate study to decide whether to use artificial substrates and, if so, how best to use them in light of study objectives.

Definitions

An artificial substrate sampler is an item of field equipment that mimics certain features of the aquatic environment into which it is placed. The sampler is colonized by macroinvertebrates at variable rates and can be retrieved after an appropriate period of exposure (Beak et al. 1973). We will designate samplers that closely resemble the natural substrate over, on or within which they are placed as representative artificial substrates (RAS) e.g., a tray filled with rocks having a similar size distribution to the natural stream bottom and placed within the stream substratum. In contrast, standardized artificial substrates (SAS) are samplers that differ from the natural substrate of the habitat in which they are placed (e.g., a tempered hardboard multiplate sampler suspended over a stony stream substratum).

Types of Artificial Substrates

The major categories of artificial substrates are described in Chapter 7 of this book.

ADVANTAGES AND DISADVANTAGES IN THE USE OF ARTIFICIAL SUBSTRATES

A review of the literature on artificial substrates indicates several advantageous and disadvantageous features associated with their use. Those features most commonly identified, or considered by us to be important, are summarized in Tables I and II and discussed individually below.

Advantages

Advantage 1: Allow Collection of Data from Locations that Cannot Be Sampled Effectively by Other Means

Artificial substrates have been used to sample benthos in habitats for which either no appropriate sampling device is available or existing samplers are inefficient. These include habitats with hard (Hilsenhoff 1969) or shifting substrates (Crossman and Cairns 1974), large boulders (Dickson et al. 1971; Rosenberg and Wiens 1976), deep or fast flowing water (Anderson and Mason 1968; Crowe 1974) or submerged obstacles (Crossman and Cairns 1974). Artificial substrates are also useful in habitats that may be decimated by conventional sampling (Macan and Kitching 1972; Mason 1976) (see Advantage 6). In some situations artificial substrates can facilitate sampling. For example, they can be placed and retrieved during a range of weather and stream conditions (Crowe 1974; Armitage 1979). Attachment to manmade objects such as river locks provides a safe way to collect benthic macroinvertebrates from large rivers when dangerous conditions would preclude sample collection directly from boats, by scuba or by other means (Elbert 1978). We conclude that artificial substrates are useful sampling devices simply

Table I. Advantages (most commonly identified and/or important) of Using Artificial Substrates to Sample Freshwater Macroinvertebrates

Advantage	References
1. Allow collection of data from locations that cannot be sampled effectively by other means.	Hilsenhoff (1969); Cairns and Dickson (1971); Dickson et al. (1971); Disney (1972); Macan and Kitching (1972); Beak et al. (1973); Weber (1973); Crossman and Cairns (1974); Crowe (1974); APHA (1975); McConville (1975); Macan (1977); Fredeen and Spurr (1978); Rabeni and Gibbs (1978); Herrmann (1979); Meier et al. (1979); Deutsch (1980)
2. Permit standardized sampling[a] by:	Hilsenhoff (1969); Cairns and Dickson (1971); Dickson et al. (1971); Dickson and Cairns (1972); Mason et al. (1973); Slack et al. (1973); Crossman and Cairns (1974); APHA (1975); Pearson and Jones (1975); Hellawell (1978); Barber et al. (1979); Myers and Southgate (1980)
eliminating subjectivity of choice and sampling at a number of different stations,	Cairns and Dickson (1971); APHA (1975); Barber et al. (1979); Meier et al. (1979)
reducing confounding effects of differences in habitat (e.g., substrate, depth) and time, etc.	Anderson and Mason (1968); Hilsenhoff (1969); Cairns and Dickson (1971); Sladeckova and Pieczynska (1971); Beak et al. (1973); Mason et al. (1973); Slack et al. (1973); Weber (1973); Crossman and Cairns (1974); Pearson and Jones (1975); Macan (1977); Voshell and Simmons (1977); Cover and Harrel (1978); Khalaf and Tachet (1978); Roby et al. (1978); Kovalak (1979); Meier et al. (1979); Shaw and Minshall (1980); Wefring and Teed (1980)
3. Reduced variability compared with other types of sampling.[b]	Beak et al. (1973); Weber (1973); Hughes (1975); Fredeen and Spurr (1978); Khalaf and Tachet (1978); Rabeni and Gibbs (1978); Meier et al. (1979); Myers and Southgate (1980)
4. Require less operator skill than other methods and, therefore, may be usable by inexperienced field biologists (e.g., placement and collection in extensive water quality monitoring programs).	Anderson and Mason (1968); Fullner (1971); Slack et al. (1973); Pearson and Jones (1975); Hellawell (1978)

5. Convenient to use.	Dickson et al. (1971); Mason et al. (1973); Pearson and Jones (1975); Roby et al. (1978)
Samples contain negligible amounts of extraneous material (or lower amounts than with other sampling methods) and, thus, can be processed quickly (i.e., cleaned, sorted, etc.).	Hilsenhoff (1969); Macan and Kitching (1972); Weber (1973); Pearson and Jones (1975); Macan (1977); Fredeen and Spurr (1978); Khalaf and Tachet (1978); Barber et al. (1979)
Samplers (e.g., hardboard multiplates, plastic strips, etc.) are small, light, inexpensive and simple to construct (and replicate), easy to handle (anchor, retrieve, package for return to laboratory) and/or surface area is easy to calculate.	Fullner (1971); Macan and Kitching (1972); Beak et al. (1973); Mason et al. (1973); Slack et al. (1973); Crowe (1974); Macan (1977); Boobar and Granett (1978); Tsui and Breedlove (1978); Fredeen and Spurr (1978); Meier et al. (1979); Wefring and Teed (1980)
6. Permit nondestructive sampling of an "environment."	Macan and Kitching (1972); Macan (1977); Myers and Southgate (1980)
7. Permit greater flexibility in sampling programs.	Weber (1973); Fredeen and Spurr (1978); Meier et al. (1979); Wefring and Teed (1980)

[a] Also stated as: uniform and reproducible substrate composition and area permitting collection of benthos in various types of stream bottom without affecting efficiency (Mason et al. 1973; Meier et al. 1979). And: "Uniformity is assured, the area or volume sampled is constant and one might, therefore, reasonably expect the results to be reproducible and comparable" (Hellawell 1978, p. 58).

[b] Also stated as "a higher level of precision is obtained than with other sampling devices" (Cover and Harrel 1978, p. 81).

Table II. Disadvantages (most commonly identified and/or important) of Using Artificial Substrates to Sample Freshwater Macroinvertebrates

Disadvantage	References
1. Colonization dynamics of artificial substrates are incompletely known.	Disney (1972); Mason et al. (1973)[a]; Sheldon (1977)
May be selective for standing crop (numbers, biomass) and/or kinds of organisms that colonize them.[b]	Sladeckova and Pieczynska (1971); Disney (1972); Glime and Clemons (1972); Macan and Kitching (1972); Slack et al. (1973); Weber (1973); Benfield et al. (1974); Crossman and Cairns (1974); Hughes (1975); Macan (1977); Minshall and Minshall (1977); Cover and Harrel (1978); Fredeen and Spurr (1978); Hellawell (1978); Khalaf and Tachet (1978, 1980); Rabeni and Gibbs (1978); Roby et al. (1978); Barber et al. (1979)
Duration of exposure required to achieve optimal colonization is often unknown.	Dickson and Cairns (1972); Pearson and Jones (1975); Fredeen and Spurr (1978)
Colonization varies seasonally.	Anderson and Mason (1968); Macan and Kitching (1972); Mason et al. (1973); Mason (1976); Freeden and Spurr (1978); Deutsch (1980); Shaw and Minshall (1980)
2. Artificial substrates do not provide a measure of the condition of the natural substrate and/or the effect of pollution on that substrate.	Weber (1973); Hughes (1975); Rabeni and Gibbs (1978)
3. Require relatively long exposure time to obtain a sample of equilibrium populations/communities.	Mundie (1956); Sladeckova and Pieczynska (1971); Hellawell (1978)
Long colonization time precludes prompt study of possibly important environmental changes during that time.	Mason (1976)

Makes artificial substrates unsuitable for short-term surveys/monitoring.	Weber (1973); Khalf and Tachet (1978)
Increases possibility that samplers will be lost.	Sheldon (1977)
4. Loss of fauna on retrieval.	Beak et al. (1973); Slack et al. (1973); Crowe (1974); Hellawell (1978); Khalaf and Tachet (1978); Rabeni and Gibbs (1978)
5. Subject to loss (and/or altered effectiveness) by spate, drought, vandalism, or burial through sedimentation.	Hilsenhoff (1969); Weber (1973); Crowe (1974); Fredeen and Spurr (1978); Hellawell (1978); Khalaf and Tachet (1978); Roby et al. (1978)
6. Inconvenient to use and logistically awkward.	
Two visits are necessary to take one sample (i.e., setting and recovery).	Hellawell (1978); Rabeni and Gibbs (1978); Roby et al. (1978)
Handling problems:	
a) Samplers can be cumbersome, heavy, and awkward to store and transport.	Fullner (1971); Beak et al. (1973); Roby et al. (1978)
b) Installation can be difficult (e.g., anchoring) and may require appreciable time. Samplers may be navigation hazards.	Weber (1973); McConville (1975); Mason (1976); Hellawell (1978)
c) Samplers may require appreciable time and effort to find, recover and clean. Cleaning usually must be done in the field. Samples may be large, requiring time to process (i.e., sieving, sorting, etc.)	Fullner (1971); Beak et al. (1973); Stanford and Reed (1974); Mason (1976); Voshell and Simmons (1977); Fredeen and Spurr (1978)

[a]Also stated as: "Factors affecting their performance. . .have not been thoroughly evaluated" (Mason et al. 1973, p. 410).

[b]Also stated as: Artificial substrates may not provide an accurate assessment of standing crops or relative abundance and relative biomass of predominant groups found on the adjacent stream bed (cf. Minshall and Minshall 1977).

because they are the only type of sampler that can be used in certain situations.

Advantage 2: Permit Standardized Sampling

One of the most commonly claimed advantages of using artificial substrates is that they permit standardized sampling (Table I). For example, Cairns and Dickson (1971, p. 760) reported that artificial substrates "to some extent, standardize the sampling procedure." The implications of this or similar statements can be interpreted as follows:

1. Subjectivity in choice of habitat to be sampled is eliminated.
2. Subjectivity in actually taking the sample is eliminated.
3. Confounding effects of differences in habitat are eliminated.

Subjectivity in Choice of Habitat. This is really a consideration of *macrohabitats* (e.g., riffles, pools, etc.) as opposed to *microhabitats* (e.g., different substrate particle size distributions, etc.), which are discussed below in the section entitled Confounding Effects of Differences in Habitat. This advantage implies that the choice of macrohabitat in the use of direct sampling methods is important, whereas the choice of macrohabitat into which artificial substrates are to be placed is not important. In actuality, a choice of habitat must be made in either type of sampling. Placement of the same type of artificial substrate in two different habitats is not standardization. Cairns and Dickson (1971, p. 760) indicated this when advising that "artificial substrates must be placed in ecologically similar conditions. . . ." Unfortunately, the literature contains many examples of the simultaneous use of artificial substrates in a variety of macrohabitats. Hilsenhoff (1969) attempted to demonstrate sampler versatility by using a single type of artificial substrate to compare populations in headwater sites with moderate current and sand and silt substrates; very fast mid-reach sites with rock and gravel substrates; and a slow wide downstream site with a sand, fine gravel and silt substrate. However, large-scale surveys frequently use a single artificial sampler type to sample and compare a range of habitat types (e.g., Anderson and Mason 1968; Mason et al. 1970).

Although the range of macrohabitats that can be encompassed within the "ecologically similar conditions" description of Cairns and Dickson (1971) remains to be defined, we conclude that the use of artificial substrates in different macrohabitats cannot serve to standardize a sampling program.

Subjectivity in Taking Samples. A convincing argument can be made that the overall role of the operator is less important in taking artificial substrate samples than in direct sampling methods. The setting and retrieval of artificial substrates may be subject to operator error, but the actual sampling is done through colonization of the artificial substrate and, therefore, is independent of the operator. All aspects of direct sampling methods (e.g., Surber sampler, Ekman grab) are influenced by the operator, who must expend the same effort and degree of efficiency for each sample to ensure standardization. Hellawell's (1978) analogy that artificial substrates are "passive" samplers, i.e., self-collecting, whereas direct sampling methods are "active" is appropriate. We conclude that this is a true advantage in standardization.

Confounding Effects of Differences in Habitat. This advantage assumes that *microhabitats* offered to benthic macroinvertebrates by replicates of a single artificial substrate type are the same; therefore, colonization by macroinvertebrates will be the same on each replicate and standardized sampling is achieved. The logic has been variously stated; for example, Beak et al. (1973) concluded that difficulties in the evaluation of site differences in pollution surveys of macroinvertebrates will result unless variations in bottom conditions are eliminated through the use of artificial substrates. Barber et al. (1979) noted that, in contrast to naturally occurring conditions, artificial substrates offer the same (micro)habitat at all sampling stations (see also Myers and Southgate 1980).

The argument presented depends on an assumption that replicate artificial substrates are placed into similar macrohabitats (see above) and on an ability to standardize replicates of a single type of artificial substrate. The features that are involved in this standardization include, but certainly are not limited to, surface area (e.g., Minshall and Minshall 1977) (see discussion below), volume (e.g., Khalaf and Tachet 1980), amount of interstitial space (e.g., Reice 1980), size and shape of contained particles (e.g., Hart 1978), texture (e.g., Markosova 1980), and particle size distribution (e.g., Allan 1975). Surface area usually receives the most emphasis in artificial substrate studies and is considered in detail here. The importance of standardizing other features has received less attention and is less well understood (e.g., Reice 1980).

The confounding effects of comparing samples with different surface areas, a problem of direct sampling methods, presumably is eliminated by using replicates of artificial substrates having the same surface areas and placed at different sites. Reasonably close agreement between RAS and direct stream bed sampling may be obtained if material identical in composition and packing to that on the stream bed is used in the artificial substrate sampler (Minshall and Minshall 1977). Resh (1979) suggested that a major limitation in current benthic studies is an inability to consistently take samples of known unit, especially with regard to the surface area of contained substrate particles. When the density of organisms is examined with reference to the area or volume of the substrate rather than to the area actually available in terms of stone surface, mean density estimates will have a variance with two components: (1) degree of population aggregation, and (2) each sample may contain different amounts of stone surface (Calow 1972; Resh 1979). Inequality of exposed surface area has been related to different results in comparisons among artificial substrates (e.g., Mason et al. 1973) and between artificial substrates and other sampling methods (e.g., Roby et al. 1978).

It is easy to standardize surface area for most types of SAS, such as porcelain balls (e.g., Jacobi 1971, and Roby et al. 1978); frustums (truncated concrete cones) (Elbert 1978); and various types of floating SAS (Fredeen and Spurr 1978). Two-dimensional SAS, such as multiplates, tiles, etc., require simple linear measurements. Similarly, replicates of artificial vegetation can be constructed to yield uniform size and surface area (e.g., Barber et al. 1979). Surface area of natural substrate in RAS can be determined using methods described by Calow (1972), Minshall and Minshall (1977) and Reice (1980), for example, but obviously with greater difficulty than for most SAS. The standardization of more complex artificial substrates (e.g., RAS) is complicated by interactions among the features mentioned above (e.g., Reice 1980).

A number of warnings should be mentioned here:

1. The initial known surface area of an artificial substrate may change with use. For example, with increasing exposure time the accumulation of fine sediment and organic matter in interstitial spaces can alter predetermined surface area (cf. Roby et al. 1978). If surface area is not radically altered by exposure, then the known surface area feature is undoubtedly a distinct advantage in using artificial substrates.
2. Use of SAS may offer different microhabitats than are found in the natural substrate and, consequently, selectivity may result (e.g., Rabeni and Gibbs 1978). This aspect is discussed more fully under Disadvantage 1 and should be considered in evaluating the applicability of the standardization advantage to specific experimental designs.
3. Standardization can be affected by differences in depth of placement (Mason et al. 1973; Gersabeck and Merritt 1979) (see also Disadvantage 2, length of exposure time (see Disadvantage 1), time of year of initial

placement (see Disadvantage 1) and other features involving installation of artificial substrates in a macrohabitat (see discussion above).

In conclusion, our ability to standardize features of the microhabitat offered to colonizing macroinvertebrates by replicates of an artificial substrate must be considered a qualified advantage. Standardization of important features such as surface area is easy to do with most simple artificial substrates (i.e., SAS), but relatively more difficult to do with more complex artificial substrates (i.e., RAS). Additionally, interaction of substrate characteristics (e.g., particle size distribution, detritus accumulation) that would be present in complex artificial substrates is not clearly understood. However, the potential for control over features of the microhabitat has made artificial substrates especially valuable for experimental manipulations in the field (Sladeckova and Pieczynska 1971; Minshall and Minshall 1977; Myers and Southgate 1980; Shaw and Minshall 1980).

Summary and Recommendation. Critical evaluation of the claim that artificial substrates permit standardized sampling (Table I) reveals the following:

1. The choice of *macrohabitat* for sampling is as important for artificial substrates as it is for direct sampling methods.
2. Variability in operator efficiency is reduced when an artificial substrate sample is taken because the operator is not directly involved in actually collecting the sample.
3. Features of the *microhabitat* presented to benthic macroinvertebrates potentially can be standardized, but efficacy of this standardization depends on complexity of the artificial substrate involved and condition of the habitat into which the sampler is placed.

We recommend that future references to the standardization advantage critically assess how the aspects discussed above relate both to the choice of a particular artificial substrate type and to study objectives.

Advantage 3: Reduced Variability Compared with Other Types of Sampling

Of all the advantages discussed for artificial substrates, none has been as vigorously defended or deprecated as whether a higher level of precision is obtained with artificial substrates than with direct sampling devices. For example, Beak et al. (1973) and Weber (1973) indicate that SAS significantly reduced sampling variability (i.e., as a function of the number of samples necessary to yield a standard error of the mean within a given percent of the mean), as did Shaw and Minshall (1980) for RAS. However, Mason (1976), using RAS, and Roby et al. (1978) and Hellawell (1978), using SAS, reported that reduction in variability was not great. Rabeni and Gibbs (1978) reported that coefficients of variation (CV) for numbers of taxa and numbers of individuals were lower at two polluted sites for rock-filled basket samples than for Hess samples, but that CVs were similar for the two types of samples at pristine and enriched sites (see also Hughes 1975, Table 5, for numbers of individuals). However, data presented by Hughes (1975, Table 5, for numbers of species collected at a station affected by heavy sedimentation) indicate that relatively lower CVs for artificial substrate samples at polluted sites may not always occur.

Part of the controversy results from incorrect figures reported in two often-cited papers. Roby et al. (1978, p. 7), among others, presented the following information: "Coefficients of variation of total numbers varies widely in reported artificial substrate studies. The CV ranged from 0.109 (Dickson et al., 1971, limestone-filled baskets) to 0.849 (Dickson and Cairns, 1972, plastic webbing)." However, these values should be corrected to read 0.20 and 0.46, respectively, because of calculation errors made by Dickson et al. (1971, Table 2) and Dickson and Cairns (1972, Table 4). Since both papers are cited frequently, the corrected values are given in Table III.

Additionally, Hellawell (1978) was critical both of the method used by Dickson et al. (1971) to compare results from their basket samplers with Surber sampler results from Needham and Usinger (1956) and of the conclusion of Dickson et al. (1971) regarding numbers of sample replicates necessary to achieve a desired precision. Hellawell's (1978) comparison of the efficiency of the basket samplers of Dickson et al. (1971) and the Surber samples of Chutter and Noble (1966)—a reasonable substitution for Needham and Usinger (1956); see Chutter (1972) and Resh (1979)—indicates that more basket sampler than Surber sampler replicates are necessary for a desired precision, the opposite conclusion to the one usually drawn from Dickson et al. (1971) by others (e.g., Beak et al. 1973).

To circumvent problems of interpretation, we have relied on actual data rather than conclusions from previous studies. Several trends are apparent if CVs (for total numbers) from selected studies in which direct sampling devices were used are compared with those obtained using SAS (Figures 2a,b). First, both types of methods can have high or low variability. Second, one-half of the direct benthic sampling device CVs are 50% or less, whereas one-half of the SAS CVs are 30% or less. This result

Table III. Corrections to (A) Table 2 of Dickson et al. (1971) and (B) Tables 2, 4 and 5 of Dickson and Cairns (1972)

A. Dickson et al. (1971)

Station	Coefficient of Variation (%)	
	Original Value	Corrected Value
1	10.9	19.7
2	28.3	34.2

B. Corrected values for Dickson and Cairns (1972) (SD = standard deviation; CV = coefficient of variation).

Day of Sampling	Table 2		Table 4		Table 5	
	SD	CV(%)	SD	CV(%)	SD	CV(%)
7	1.73	13.3	161.4	49.9	0.21	8.0
14	0	0	186.3	36.0	0.29	11.7
21	1.73	11.5	256.1	31.3	0.13	5.2
28	3.05	18.3	1817.1	100.2	0.09	4.2
35	7.37	34.0	399.0	19.0	0.13	4.9
42	1.53	10.0	83.1	8.0	0.06	2.4
49	2.65	16.5	371.4	34.4	0.13	6.3
56	3.79	26.4	975.0	83.6	0.86	46.7
63	4.00	22.2	399.3	52.5	0.50	20.1

suggests that even with the greater variety of sampling devices and the broader range of habitats for SAS than for direct sampling methods (of the selected studies used in Figures 2a,b), use of SAS may result in at least a 10%, if not a full 20%, reduction in CVs. Less information is currently available on RAS, but analysis of RAS CVs suggests a greater similarity to SAS CVs than to those of direct sampling approaches (Figure 2c). There are two possible reasons: (1) uncorrected mechanical bias in direct substrate sampling (Resh 1979, Table 1) are eliminated in RAS; and (2) even though RAS are representative of the natural substrate, they are more similar from replicate to replicate (e.g., due to less microhabitat patchiness) than are samples taken directly from the substrate.

To demonstrate the effect of a reduction of CV on sampling effort, two hypothetical populations with $\bar{x}$ densities = 100 and 1000 individuals, respectively, are compared over a range of CVs (Table IV). A CV reduction from 50% to 40%, for example, reduces from 7 to 4 the number of sample replicates needed to obtain an estimate of the population mean (±40% of the mean; 95% confidence limits).

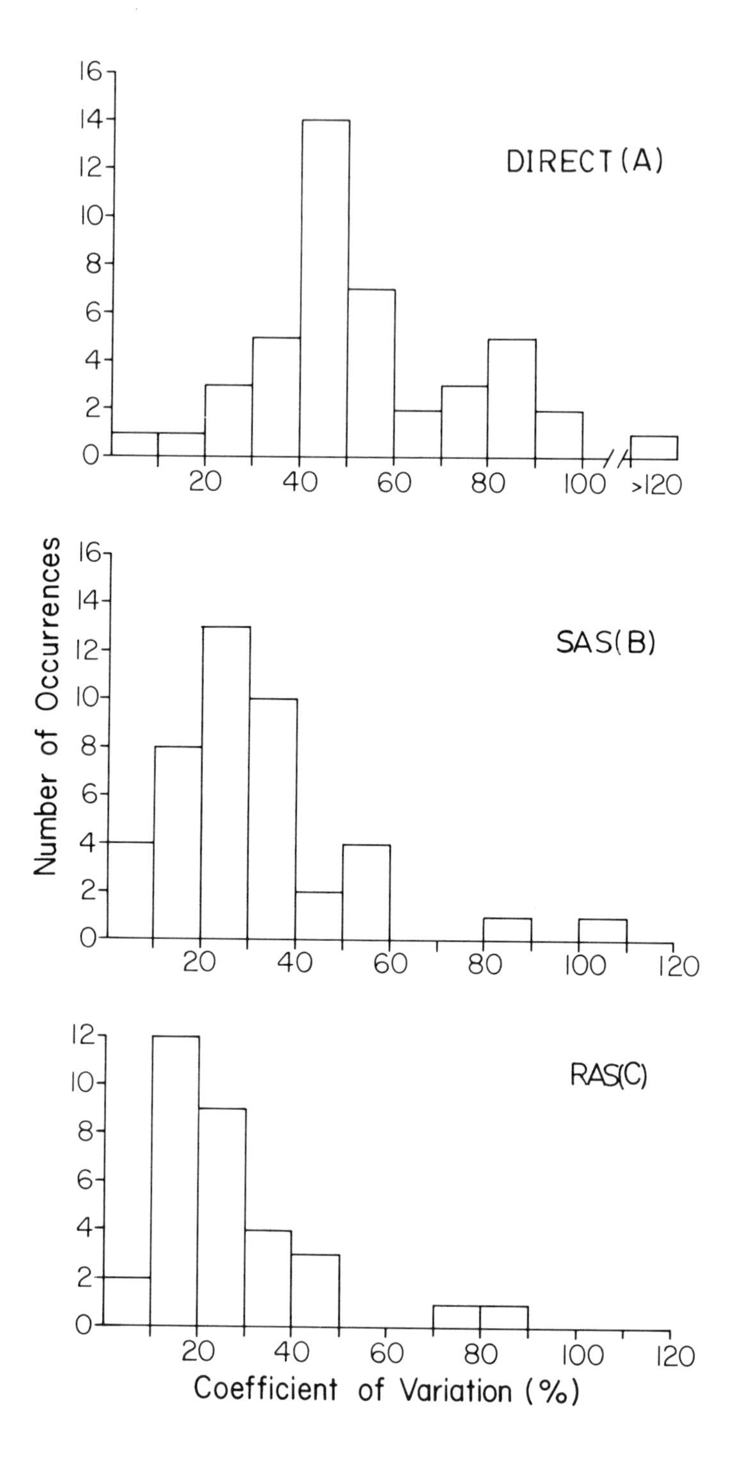
DIRECT (A)
SAS (B)
RAS (C)
Number of Occurrences
Coefficient of Variation (%)
0
2
4
6
8
10
12
14
16
20
40
60
80
100
>120
120

Table IV. Coefficients of Variation (CV) and Required Sample Sizes (N) (±40% of the mean; 95% confidence limits) for Mean Densities of 100 (A) and 1000 (B) Individuals (calculations according to Elliott 1977; N given to nearest integer)

Mean Densities ± SD		CV(%)	N
A	B		
100±20	1000±200	20	1
100±30	1000±300	30	3
100±40	1000±400	40	4
100±50	1000±500	50	7
100±60	1000±600	60	9

Figure 2. A: Coefficients of variation (n=44) for total numbers of individuals collected by direct substrate sampling methods in selected studies. Data are based on riffle habitat collections made with a Surber sampler (Needham and Usinger 1956; Allen 1959[a]; Hales 1962; Chutter and Noble 1966; Weber 1973; Hughes 1975; Hornig and Pollard 1978; Jacobi 1978; Roby et al. 1978; Lamberti and Resh 1979; Resh 1979 and unpublished data) and a Hess sampler (Mason 1976; Minshall and Minshall 1977; Shaw and Minshall 1980).
B: Coefficients of variation (n=42) for total numbers of individuals collected by standardized artificial substrate (SAS) sampling methods in selected studies. Data are based on collections made with floating baskets (Mason et al. 1973; Weber 1973; Benfield et al. 1974); floating conservation webbing (Dickson and Cairns 1972; Benfield et al. 1974); floating multiplates (Mason et al. 1973; Cover and Harrell 1978); bottom baskets in riffles (Benfield et al. 1974), pools (Roby et al. 1978), and lakes (Voshell and Simmons 1977); and bottom ceramic tiles in riffles (Lamberti and Resh, unpublished data). Material within baskets differed from study to study.
C: Coefficients of variation (n=32) for total numbers of individuals collected by representative artificial substrate (RAS) sampling methods in selected studies (Hughes 1975; Mason 1976; Khalaf and Tachet 1977; Minshall and Minshall 1977; Rabeni and Gibbs 1978; Shaw and Minshall 1980; Lamberti and Resh, unpublished data ; Resh, unpublished data). Data are based on collections made with RAS implanted in or placed on top of the natural substrate.

[a]Sampler type assumed from author's description, "1 sq. ft."

The above discussion indicates that artificial substrates generally do provide less sampling variability and increased sampling precision over direct sampling methods. Variability in benthic macroinvertebrate studies that yield high CVs may result from two components: sampling device inconsistency and spatial distribution of the macroinvertebrates. The lower sampling variability of artificial substrates may reflect a reduction in sampling device inconsistency, a change in macroinvertebrate spatial distribution, or both. A nonrandom distribution pattern would tend to show a large intersample variation even if the sampler were consistently efficient in capturing the animals in the area sampled (Hughes 1975; Resh 1979). Nonrandom distribution patterns typically are reported for benthic macroinvertebrate populations (Resh 1979), and Shaw and Minshall (1980) reported these patterns for selected populations on RAS. The spatial arrangement patterns of some benthic macroinvertebrates (e.g., blackflies) probably would be the same on artificial substrates (e.g., polyethylene strips) as on natural substrates (e.g., flat bedrock surfaces). However, for most other benthic macroinvertebrates it is not known whether the spatial arrangement of a population differs from artificial to natural substrates.

We conclude that artificial substrates generally require fewer sample replicates than direct sampling devices to achieve a given sampling precision. However, the word "precision," as used above, refers to a statistical reduction in sampling variability, but not necessarily an increase in sampling "accuracy" (i.e., the true representation of conditions that exist in the habitat being studied). Information potentially gained from knowledge of spatial arrangements of individuals may be sacrificed in achieving greater sampling precision. The objective of a study will determine the importance of this tradeoff.

Advantage 4: Require Less Operator Skill than Other Methods

The placement and retrieval of artificial substrates by nonbiologists has been claimed by a number of authors as an advantage over other sampling methods (Table I). Long-term surveys in which many sampling stations are operated over a large area would benefit from such involvement. The economic advantages are obvious. Labor costs could be reduced by hiring nonbiologists or by involving cooperators such as forest rangers who are located in the area of a sampling program.

However, problems of involving nonspecialists in a sampling program also must be considered. For example, some artificial substrates are easier to install or retrieve than others. More complex artificial substrates, especially those requiring field separation of organisms from the sampler, may pose difficulties for the nonspecialist and result in greater variability in the data. An attempt to avoid the problem by using only simple artificial substrates solely to capitalize on this advantage could greatly limit sampling design.

A system involving nonbiologists in the collection of biological data undoubtedly requires some onsite training and an opportunity for communication where questions can be answered and departures from standard operating practices approved and recorded. The choice of particular habitats in which artificial substrates are placed should not be left to an untrained person.

We have been unable to determine the extent of involvement of nonbiologists in sampling programs using artificial substrates, but we conclude that such involvement is a potential advantage in the economics of large-scale programs, as long as the drawbacks listed above are considered as well.

Advantage 5: Convenient to Use

There are several aspects to the claim that artificial substrates are convenient to use (Table I):

1. Artificial substrates collect less debris than other sampling methods, making samples easier to clean and sort.
2. Artificial substrates are small, light, inexpensive and simple to construct.
3. Artificial substrates are easy to handle.
4. Surface area is relatively easy to calculate.

Ease of Processing. No matter how effective the experimental design of a study, sampling bias will result if organisms are not removed completely from the accompanying debris in a sample. The sorting of benthic macroinvertebrates is a labor-intensive activity that greatly increases the cost of freshwater macroinvertebrate studies; therefore, samples that contain negligible amounts of extraneous material and can be rapidly cleaned and sorted provide obvious benefits.

Our examination of the literature revealed conflicting evidence for the claim that artificial substrate samples are easier to process than those of other sampling methods. Several authors indicate that artificial substrates are relatively easy to clean and that samples can be sorted quickly (e.g., Hilsenhoff 1969; Lewis and Bennett 1974; Macan 1977; Barber et al. 1979). However, debris accumulation in some locations can increase sorting time (e.g., Zillich 1967; Fredeen and Spurr 1978; Roby et al. 1978). Sorting time also may vary with type (Crowe 1974) or size (Voshell and Simmons 1977) of artificial substrate, and may be greater for artificial substrates than for direct sampling methods (e.g., Rabeni and Gibbs 1978); (see also Disadvantage 6).

Simple artificial substrates (e.g., polyethylene strips and tiles) usually will require less sorting time than directly-taken benthic samples, although site-specific factors (e.g., growth of filamentous algae) may increase sorting time. As artificial substrates become more structurally complex, a consistent reduction in sorting time between these and direct sampling methods becomes less predictable. We conclude that artificial substrate samples are not always easy to process and that site-specific factors can make processing more difficult even for those artificial substrates that are structurally simple.

Small, Light, Inexpensive and Simple to Construct. It is impossible to generalize about size and weight of artificial substrates because these features will depend on the type of artificial substrate being considered. Most of the artificial substrate samplers listed in Table I of Chapter 7 are constructed relatively simply, far more so than most other samplers (Merritt et al. 1978; Figures 3.6–3.9, 3.11–3.19 vs Figures 3.20–3.22). Because most artificial substrates are simple and inexpensive to build, only selected types can be purchased from biological supply houses (Merritt et al. 1978, Table 3A). Artificial substrates such as polyethylene strips (e.g., Fredeen and Spurr 1978; Gersabeck and Merritt 1979), leaves (e.g., Disney 1972) or leaf packs (e.g., Merritt et al. 1979, Figure 2) are less expensive than porcelain balls (e.g., Jacobi 1971; Roby et al. 1978) or the implanted colonization boxes used by Peckarsky (1979, 1980). A need for replicates and the one-time-only use aspect of many artificial substrates will raise their cost, but other samplers are unlikely to be chosen because they are less expensive than artificial substrates. In the final analysis, sampling equipment costs are usually only a small fraction of total costs of most relatively large freshwater macroinvertebrate studies.

We conclude that ease of construction and low cost are definite advantages of using artificial substrates and are factors that will be appreciated most in small-scale studies.

Ease of Handling. This is a subjective appraisal reported by several investigators. Features of the sampler such as its construction, size and weight, the ease with which it can be placed in the habitat (see Disadvantage 6), the habitat into which it is placed (e.g., a large turbulent river vs a small stream), and seasonal or even ambient weather conditions will greatly affect an individual operator's judgement of handling ease. In any event we feel strongly that ease of handling, like the advantage of requiring less operator skill, should not be considered an overriding advantage in the choice of a particular artificial substrate type.

Ease of Calculating Surface Area. This advantage also will depend on the type of artificial substrate under consideration. Surface areas of devices such as ceramic tiles and polyethylene strips, can be calculated easily, whereas baskets filled with rocks of varying sizes present greater difficulties. A more thorough consideration of surface area appears under Advantage 2.

Summary. We conclude that all aspects of the convenience of using artificial substrates, except their low cost and generally simple design, must be qualified by the type of artificial substrate being considered and the conditions under which it will be used.

Advantage 6: Permit Nondestructive Sampling of an Environment

Regular sampling of a small habitat without harming it was cited as an advantage of using artificial substrates by Macan and Kitching (1972) and Macan (1977) in work done on Hodson's Tarn, a small English pond. Myers and Southgate (1980) recommended the use of artificial substrates (nylon pan-scourers) to avoid destructive sampling of naturally occurring red algal turfs during a monitoring program for cryptofauna on rocky shores. Habitat destruction by sampling in a small stream was also a concern expressed by Mason (1976).

There is no doubt that reducing habitat destruction is an advantage of sampling by artificial substrate. However, redistribution and, hence, lower densities of macroinvertebrates, will result following sampler removal unless sampling is done during a period of population recruitment (e.g., oviposition). Lamberti and Resh (1979), in a discussion of the effects of destructive sampling on density estimates of benthic populations in confined habitats, showed how population redistribution would result in lower densities simply as an artifact of sampling. They described how behavioral and phenological information can be incorporated into sampling designs to avoid this problem. Studies of rare populations or those in unusual habitats often must be designed to avoid population reduction. Unless a census rather than a removal method of density estimation is used, artificial substrates will be no better than direct sampling methods in this regard. For example, Lamberti and Resh (unpublished data) devised a photographic method for counting stream macroinvertebrates colonizing ceramic tiles.

Advantage 7: Permit Greater Flexibility in Sampling Programs

The intended meaning of this advantage (Table I) is unclear. The following different interpretations are possible:

1. Artificial substrates are mechanically pliable (*sensu* Bull 1968; Armitage 1979), an unlikely connotation!
2. Basic sampler designs can be modified to suit specific situations (e.g., see Chapter 7, Table I). However, the result of this activity is to make comparisons of data from different studies difficult because so many different materials, shapes, placement techniques, etc., have been used (e.g., Gersabeck and Merritt 1979). The solution is standardization of sampler type (e.g., Lewis and Bennett 1974), but then flexibility in this sense is eliminated as an advantage.
3. Artificial substrates can be used in a wide range of habitats (Chapter 7, Table I) and/or in habitats where other samplers either cannot be used or are inefficient (see Advantage 1). This is a valid point.
4. A single type of artificial substrate could be used to sample a range of habitats. This is also a valid point, but the cautions regarding use of samplers in ecologically similar habitats discussed under the section on standardization (see Advantage 2) and considerations of selectivity (see Disadvantage 1) would apply to this interpretation.

We conclude that in future each author who uses the term "flexibility" should clearly define its meaning. In addition, referring to this advantage in citing past work should be avoided.

Disadvantages

Disadvantage 1: Colonization Dynamics Incompletely Known

The lack of knowledge about colonization dynamics of artificial substrates appears to be the most serious disadvantage in their use. Factors affecting the colonization and distribution of aquatic invertebrates generally appear to be complex and difficult to study (e.g., Rabeni and Minshall 1977; Peckarsky 1980; Shaw and Minshall 1980). Fortunately, this aspect of freshwater invertebrate ecology has received increased attention recently (e.g., Ulfstrand et al. 1974; Townsend and Hildrew 1976; Williams and Hynes 1976a, b; Rabeni and Minshall 1977; Sheldon 1977; Williams 1977; Peckarsky 1979, 1980; Shaw and Minshall 1980; Williams 1980). In addition to a general lack of detailed studies dealing with colonization processes of artificial substrates, three aspects of colonization dynamics seem particularly troublesome (Table II): (1) selectivity (i.e., nonrepresentativeness) of artificial substrates compared to the natural substrate; (2) duration of exposure of artificial substrates necessary to achieve optimal colonization (i.e., equilibrium levels of whatever population measurement is used); and (3) seasonal variations in data collected.

Selectivity. Selectivity refers to differences in the relative abundances of populations occurring on artificial substrates compared to the natural substrate. Some authors have reported relatively good correlations in fauna sampled between artificial and natural substrates (e.g., Pearson and Jones 1975; Herrmann 1979), but most users of artificial substrates have reported them to be selective (Table II). In examining selectivity, emphasis will be placed on whether colonization mechanisms on artificial substrates result in a fauna that is representative of the habitat under study.

1. Factors in Colonization of Artificial Substrates. A number of factors are involved in the colonization of artificial substrates. Obviously, a potential colonist must first contact the new habitat; therefore, placement of the artificial substrate is extremely important. For example, floating artificial substrates can be colonized only by species that leave the substrate and enter the water column; a drift mechanism or active swimming must be assumed. Therefore, floating artificial substrates will lack taxa that do not at least enter the water column during some phase of their life cycle (e.g., Anderson and Mason 1968; Hilsenhoff 1969; Bournaud et al. 1978; Elbert 1978) (see also Disadvantage 2). Once individual macroinvertebrates contact the new artificial substrate, numbers that eventually colonize it are a function of several features (Disney 1972): (1) density of the source population; (2) relative abundance of other suitable substrata; (3) length of time the artificial substrate is exposed; (4) intensity and nature of the factors causing the individual to move from its previous location; and (5) acceptability of the artificial substrate. Additionally, Mason et al. (1973) and Gersabeck and Merritt (1979) demonstrated that numbers of macroinvertebrates colonizing floating artificial substrates depend on depth of placement of the sampler in the water column.

2. Evidence of Selectivity. Data from studies that included a variety of artificial substrate types were examined for evidence of selectivity. Table V shows that artificial substrates are frequently selective for certain taxa when compared to methods that have sampled the substrate directly. There are probably three main reasons involved. First, separation of the artificial substrate from the habitat being directly sampled (e.g., a floating RAS vs placing it on the bottom) will lead to an altered represen-

Table V. Partial Listing of Studies in which Authors Presented Data or Commented on Selectivity of Artifical Substrates

Reference	Type of Artificial Substrate(s)	Selectivity
Anderson and Mason (1968)	Floating limestone-filled baskets.	Organisms occurring in bottom sediments underrepresented.
Hilsenhoff (1969)	Suspended cylinders with wire mesh.	Sediment-inhabiting organisms not collected.
Dickson et al. (1971)	"Bottom" basket filled with limestone chips.	Mayflies, snails and amphipods sometimes collected in disproportionately high numbers.
Jacobi (1971)	Baskets filled with spheres.	11 of 16 taxa common to both natural and artificial substrates; same species dominant on both.
Dickson and Cairns (1972)	Conservation webbing.	"selectively sample... mayflies, stonefiles, hellgrammites and caddisflies" (p. 68).
Disney (1972)	Standardized artificial substrate units (SASUs)	Particular SASUs types selective for different *Simulium* spp.
Glime and Clemons (1972)	String and plastic "mosses."	Fewer individuals than on real moss but good correlation among species.
Macan and Kitching (1972)	Artificial *Carex* and *Littorella*.	Animals more numerous in artificial vegetation.
Mason et al. (1973)	Floating limestone-filled baskets and hardboard multiplates.	Preference shown by individual species for one or the other sampler type.
Benfield et al. (1974)	Conservation webbing; concrete cones.	Selective for netspinning Trichoptera and Diptera.
Crossman and Cairns (1974)	Floating samplers (conservation webbing and styrofoam).	Selective for beetles, mayflies and caddisflies.
	"Bottom" basket filled with rocks and leaf detritus.	"collect a more representative sample than the floating samplers when compared with conventional sampling methods" (p. 520).
Hughes (1975)	Substratum-filled trays.	Higher numbers of Trichoptera, Diptera,

Table V, continued

		Gastropoda, Crustacea and Hirudinea in trays at various sites.
Mason (1976)	Substrate tray.	Positive or negative bias by over half of the most abundant taxa.
Macan (1977)	Artificial *Littorella*.	Some species not well attracted; many species more numerous in real than artificial vegetation.
Minshall and Minshall (1977)	Substratum-filled trays.	Numbers and kinds of organisms in uniform-sized substrate trays different from natural stream bed Hess samples.
Voshell and Simmons (1977)	Baskets filled with conservation webbing, leaves or limestone.	Selectively colonized by amphipods, gastropods and chironomids; no sediment inhabiting organisms in rock-filled baskets placed on lake bottom.
Bournaud et al. (1978)	Floating limestone-filled baskets.	Different fauna in floating baskets than ones on bottom.
Cover and Harrel (1978)	Hardboard multiplates.	Species composition differed from benthos.
Elbert (1978)	Floating frustum-filled baskets.	Burrowing mayflies collected in low numbers.
Fredeen and Spurr (1978)	Various types of floating standardized artificial substrates (SAS).	Odonates, hemipterans and other taxa rarely attached to SAS.
Rabeni and Gibbs (1978)	"Bottom" rock-filled baskets.	Biased for some macroinvertebrate populations.
Roby et al. (1978)	Porcelain balls in baskets.	"samplers. . .collected some organisms in greater proportion than their occurrences in the natural substrate" (p. 1).
Barber et al. (1979)	Artificial seagrass.	Burrowing forms were underrepresented; more

Table V, continued

		mobile taxa were overrepresented.
Gersabeck and Merritt (1979)	Ceramic tiles and polyethylene strips.	Individual species of blackflies showed preference for attachment materials.
Reice (1980)	Leaf packs and rock-filled baskets.	Strong preference by common taxa for one or the other sampler type.
Shaw and Minshall (1980)	Substratum-filled trays.	Underrepresentation by Chironomidae and several other taxa compared to riffle Hess samples.

tation of the benthic community on the artificial substrate (e.g., Fredeen and Spurr 1978; Khalaf and Tachet 1978; Roby et al. 1978) (see also Disadvantage 2). Second, a difference in the quality and quantity of microhabitats between artificial substrates and the natural substrate will affect the type and proportion of macroinvertebrate colonizers (e.g., Hellawell 1978; Rabeni and Gibbs 1978; see also Advantage 2). Third, biological interactions that ultimately determine benthic macroinvertebrate community structure may occur differently on artificial than on natural substrates because of a variety of biotic and abiotic factors (e.g., availability of food). Glime and Clemons (1972) reported little difference in invertebrate colonization of real moss (*Fontinalis* sp.) compared to artificial moss (Table V), perhaps because moss is seldom used as food but may be used as a substrate (see also Myers and Southgate 1980). Additionally, artificial substrates may provide preferred habitats for predators such as fish (e.g., Roby et al. 1978; Rosenberg and Wiens, unpublished data). This could alter the quality and quantity of macroinvertebrates, giving the appearance of selectivity. Conversely, numbers of the predaceous caddisfly *Rhyacophila dorsalis* were similar on artificial and natural substrates in Hughes (1975). Osman (Chapter 3, this volume) examines this subject in more detail.

3. Sampling Comparisons. Several arguments are involved in resolving the problem of differing results from artificial substrate and direct sampling of the substrate. Weber (1973) maintained that objections to selectivity of artificial substrates were trivial since all currently available

sampling techniques are selective to some degree. This argument ignores the possible correction of factors responsible for sampling bias (Resh 1979, Tables 1,3).

Some authors have argued that the collection of higher standing crops (numbers and/or biomass) and numbers of species by artificial substrates compared to direct sampling methods is an advantage (more is better: Dickson et al. 1971; Stanford and Reed 1974; Voshell and Simmons 1977; Roby et al. 1978; or more is more efficient: Hughes 1975). However, the objective of sampling is to obtain as true as possible a representation of the natural condition. Artificial substrates occasionally may function as better "species collectors" than other methods (Hughes 1975), but because of the long colonization time required to yield a sample and the amount of work involved in using them, it would be inefficient to use them solely for this purpose. Manual collections in many different habitats probably are far more efficient to obtain lists of representative species.

Arguments concerning comparisons of sampling by artificial substrates and direct methods may be moot because microhabitats presented by an artificial substrate cannot duplicate exactly those available in natural substrate (e.g., Hellawell 1978; Rabeni and Gibbs 1978); hence, two different habitats are being compared (see also Advantage 2).

4. Influence of Selectivity on Data Interpretation. The central issue of selectivity as it applies to artificial substrates is whether data from them still will have applicability if artificial substrates do not provide an accurate representation of the type of fauna that would be obtained from direct sampling. For example, Beak et al. (1973) advised that artificial substrates must duplicate natural bottom conditions as closely as possible to measure specific biological parameters, but that measurement of such parameters may not be necessary for such purposes as pollution surveys. Thus, Beak et al. imply that a clear definition of the purpose of using artificial substrates is necessary, and we agree.

Pielou (1966) described light trap catches of adult insects as assemblages of organisms to be studied as a unit. Similarly, since artificial substrates (especially SAS) cannot duplicate the same results that would be obtained from direct substrate sampling (Table V, see above discussion), can they be viewed as individual habitats rather than as samples of other habitats? If so, then replicates of artificial substrates would represent replicated habitats.

As for any sampling method, caution must be taken in interpreting data from artificial substrates. We present the following hypothetical situation to show how selectivity can influence interpretation of results. If we imagine a two-site comparison in which site 1 has populations A–Z, site 2 has populations D–S, but only populations F–Q occur on artificial substrates, no differences would be observed when artificial substrates are used to sample the two sites. Similarly, no differences would be noted if site 1 has populations A–Q, site 2 has populations M–Z, but only populations M–Q are sampled by artificial substrates. Thus, comparisons along a segment of stream in a study of pollution will have to consider the five points influencing colonization of artificial substrates listed above (Disney 1972). For example, are different numbers of a taxon along a gradient only due to pollution effects, or could the relative abundance of suitable substrata differ along this gradient, yielding different macroinvertebrate densities on artificial substrates? Factors such as these must be considered in data evaluation.

5. Other Considerations. There is a need to consider other aspects of selectivity as well when evaluating the use of artificial substrates for particular studies. For example, how does selectivity change with length of exposure of the sampler (e.g., Gersabeck and Merritt 1979)? Are artificial substrates preferentially colonized by certain life stages of macroinvertebrates (e.g., McLachlan 1970; Markosova 1979)? Is there an interspecific, age class-related "founder advantage" that would favor a more motile colonizer (cf. Davies 1976; Waters and Resh 1979; Hart and Resh 1980)? Since artificial substrates are being used currently in field experiments for hypothesis testing of biotic interactions, questions such as these deserve detailed attention.

6. Conclusion. We consider the selectivity of artificial substrates to be a disadvantage because a sampling method should accurately portray the natural fauna. However, the importance of selectivity will be determined by the study objectives; in fact, selectivity may be a definite advantage in studying life histories of certain taxa (e.g., Crossman and Cairns 1974; Macan 1977; Fredeen and Spurr 1978).

Time to Reach Equilibrium Levels. A normal sequence of colonization could be expected to show an initial, relatively rapid increase to maximum numbers of individuals, numbers of species or biomass, and this would be followed by a relatively stable equilibrium level or fluctuations around an equilibrium level (Dickson and Cairns 1972; Markosova 1979). We have defined "duration of exposure required to achieve optimal colonization" (Table II) as the time necessary for the population measurement (numbers, biomass, etc.) to achieve this equilibrium level. Presumably the interpretation of results of sampling with artificial substrates is more accurate when an equilibrium level has been reached, although Sheldon (1977) has tried to use rate of colonization instead (see below). This subject is considered in detail by Osman (Chapter 3).

Is adequate information available regarding times required for populations to achieve optimal colonization? Recommended exposure times for artificial substrates reported in the literature certainly are variable [~2 weeks to several months (cf. Sladeckova and Pieczynska 1971; Mason et al. 1973; Slack et al. 1973; Weber 1973; APHA 1975; Cover and Harrel 1978; Khalaf and Tachet 1978)], but these recommended durations frequently lack an experimental basis.

To answer the above question, we examined 20 representative studies that contained time series data on the colonization of artificial substrates. Concerns about this disadvantage (Table II) appear to be well founded. Only six of the studies examined presented unequivocal data that populations reached an equilibrium level during the study period (McLachlan 1970; Nilsen and Larimore 1973; Boobar and Granett 1978; Gersabeck and Merritt 1979; Markosova 1979; Wise and Molles 1979). Six others demonstrated that equilibrium levels were not reached (Dickson and Cairns 1972; Mason et al. 1973; Lapchin 1977; Sheldon 1977; Leglize 1978; Tsui and Breedlove 1978), but it should be noted that experimental design was not optimal in some of these studies (e.g. insufficient time was allowed for colonization or too few sampling periods were used). Eight studies reported data indicating that some population measurements reached equilibrium levels, whereas others did not (Ulfstrand et al. 1974; Pearson and Jones 1975; Townsend and Hildrew 1976; Khalaf and Tachet 1977; Bournaud et al. 1978; Cover and Harrel 1978; Roby et al. 1978; Shaw and Minshall 1980).

Data on approximate times taken to reach equilibrium levels were summarized according to the population measurement used (Table VI). Density (as no./m^2 or fraction thereof, no. per artificial substrate unit, etc.) was the most common population measurement used, but there are insufficient data available for the same taxa or type of artificial substrate to reach solid conclusions about time required to reach equilibrium levels. The results of blackfly colonization on polyethylene strips were similar (~1 week to reach equilibrium densities), but additional verification is necessary. Two of the three studies on Chironomidae showed similar equilibrium times: (1) 12 days (Khalaf and Tachet 1977); (2) 15 days (Markosova 1979); and (3) 30 days (McLachlan 1970). Experimental designs and geographical locations differed greatly: the study in Czechoslovakia used plastic paddles suspended in fish ponds (Markosova 1979); the one in Africa used sticks suspended in a reservoir (McLachlan 1970); and the one in France used baskets of rocks placed on the bottom of a stream (Khalaf and Tachet 1977). Equilibrium time for density of "all taxa" varied from 9–35 days, although the three studies using rock-filled containers placed on gravel substrate provided similar times (Wise and Molles 1979: 9 days; Pearson and Jones 1975: 14 days; Khalaf and Tachet 1977: 12 days for "all taxa, excluding Simuliidae"). Equilibrium biomass levels for "all taxa" occurred between 8 and 35 days, although the results of two studies using sticks as artificial substrates were similar, despite habitat differences (McLachlan 1970: 30 days; Nilsen and Larimore 1973: 35 days). Again, these apparent similarities need further verification. Time required to reach equilibrium levels of mean number of taxa, diversity, and total number of taxa for the commonly used category "all taxa" are variable (Table VI), despite similarities in study habitat (i.e., running waters) and type of artificial substrate used (i.e., rock-filled containers, except for Cover and Harrel 1978 and Ruby et al. 1978). Furthermore, when studies are compared in which populations do and do not reach an equilibrium, it is apparent that similar population measurements and taxa do not always reach an equilibrium.

We conclude that there are three possible reasons for our inability to identify any consistency in times necessary to reach equilibrium among different population measurements:

1. Differences in habitats, such as rivers, streams, ponds and reservoirs, and locations, such as North America, Europe and Africa, of the studies examined. Markosova (1979) noted that variations in equilibrium colonization times cited in her study are probably due to differences in locality.
2. Incomparable experimental designs, including population parameters measured, types of artificial substrates used, length of exposure and sampling intervals. We summarized some aspects of experimental design in seven studies from Table VI that used similar artificial substrates (rock-filled containers). Table VII shows how elements of experimental design can still vary among studies using artificial substrates, making comparisons of data from these studies difficult.
3. Colonization processes may be stochastic, generally not resulting in an equilibrium (e.g., Simberloff 1979). In fact, demonstration of equilibrium by some colonization studies may depend on the statistical analysis used (Simberloff 1979).

Seasonal Variation. Artificial substrates sampled during cold weather periods usually are colonized by fewer individuals and species than during warm weather periods (e.g., Anderson and Mason 1968; Deutsch 1980; Shaw and Minshall 1980; Williams 1980). Also, biomass and rates of colonization are usually lower in winter than in summer (e.g., Macan and Kitching 1972; Shaw and Minshall 1980). However, the applicability of the above generalizations will depend on the species involved (e.g., Mason et al. 1973; Shaw and Minshall 1980) and, perhaps, geographic location (e.g., McLachlan 1970).

Since density, age structure, behavior, etc., of natural populations of aquatic macroinvertebrates normally vary seasonally (e.g., Hynes 1970), colonization dynamics of artificial substrates also may be expected to vary seasonally. Users of artificial substrates should expect seasonal variations to occur in studies involving long-term colonization. This apparent disadvantage can be minimized by ensuring that installation for short-term studies is done at similar times each year (although, since variations in weather from year to year will affect normal seasonal cycles, this approach may not always be a solution), or by using appropriate sampling designs that account for seasonal variability (e.g., Gersabeck and Merritt 1979). However, since variations in weather from year to year will affect normal seasonal cycles, this approach may not always be a solution.

Summary and Recommendations. We conclude that the colonization dynamics of artificial substrates are incompletely known. The fauna colonizing artificial substrates usually cannot be considered as representative of the natural habitat because of the selectivity of artificial substrates. Consistent trends in the time necessary for populations to achieve equilibrium levels cannot be identified because of variations in habitat, geographic location and experimental design of artificial substrate studies. It is also possible that colonization may not always yield an equilibrium. Seasonal variations will occur in sampling by artificial substrates, but this is a factor in other sampling methods as well.

There is a need to elucidate general principles underlying macroinvertebrate colonization of artificial substrates, including the selectivity aspect and the extent to which colonization is a stochastic process. Future research should include detailed studies on the operation of artificial substrates. Studies such as those of Markosova (1974, 1979) on sequences of colonization, including the species and life stages involved, and on the abiotic and biotic factors regulating colonization (e.g., Rabeni and Minshall 1977; Markosova 1979; Peckarsky 1980; Shaw and Minshall 1980) are required and should be encouraged. Such studies also may help to explain the selectivity of artificial substrates. Emphasis should be placed on colonization dynamics of specific populations (Minshall and Minshall 1977; Shaw and Minshall 1980). Also, a research program that eliminates sources of variance due to experimental design might be undertaken to compare the simultaneous operation of artificial substrates in different geographic locations, e.g., throughout North America and Western Europe. One or more government agencies would appropriately lead such a study since artificial substrates are used extensively for water quality monitoring, and an increased understanding of the operation of

Table VI. Summary of Approximate Times Necessary to Reach Equilibrium Levels in Studies Using Artificial Substrates

Population Measurement	Taxon	Reference	Total Exposure Time (and Sampling Intervals) (days)	Approximate Time to Reach Equilibrium Level (days)	Artificial Substrate Used	Habitat
Density (no./m^2, no. per sampling unit, etc.)	Simultiidae (mixture of 5 species)	Boobar and Granett (1978)	29 (3,6,10,14,21,29)	6	Polyethylene strips	River
	Simuliidae (*Prosimulium mixtum/fuscum* and *Simulium dacotensis*)	Gersabeck and Merritt (1979)	11 (0.5,0.5,1,2,3,5 7,9,11)	7	Polyethylene strips	Creek
	Chironomidae	McLachlan (1970)	120 (0.5,1,2,8,16, 30,60,120)	30	Sticks	Reservoir
	Chironomidae	Markosova (1979)	75 (5,10,15,20,25,30, 35,45,75)	15	Plastic paddles	Fish pond
	Chironomini	Khalaf and Tachet (1977)	28 (2,4,6,8,12,16,20, 24,28)	12	Baskets with rocks	Stream
	All taxa (excluding Simuliidae)	Khalaf and Tachet (1977)	Same	12	Baskets with rocks	Stream
	Baetis intermedius	Shaw and Minshall (1980)	64 (1,2,4,8,16,32,64)	4	Trays with gravel	Creek
	Capnia	Shaw and Minshall (1980)	Same	16	Trays with gravel	Creek

	Nemoura	Shaw and Minshall (1980)	Same	16	Trays with gravel	Creek
	All taxa	Nilsen and Larimore (1973)	42 (7,14,21,28,35,42)	35	Sticks	River
	All taxa	Pearson and Jones (1975)	90 (4,7,14,20,40,90)[a]	14	Boxes with gravel	River
			90 (4,8,15,20,40,90)[b]	15	Boxes with gravel	River
			200 (4,8,15,20,40,200)[c]	20	Boxes with gravel	River
	Gammarus pulex	Pearson and Jones (1975)	See [a] above	14	Boxes with gravel	River
	All taxa	Wise and Molles (1979)	19 (1,2,5,9,19)	9	Baskets with gravel	River
Biomass (wt./ m^2, wt. per sampling unit, etc.)	All taxa	McLachlan (1970)	See above	30[d]	Sticks	Reservoir
	Chironomidae	Markosova (1979)	See above	15[e]	Plastic paddles	Fish pond
	B. intermedius	Shaw and Minshall (1980)	See above	4[f]	Trays with rocks	Creek
	Capnia	Shaw and Minshall (1980)	See above	8	Trays with rocks	Creek
	All taxa	Shaw and Minshall (1980)	See above	8	Trays with rocks	Creek
	All taxa	Nilsen and Larimore (1973)	See above	35[e]	Sticks	River
Mean No. Taxa, Species Per Sampling Unit	All taxa	Roby et al. (1978)	70 (weekly for 10 weeks)	28	Baskets with porcelain balls	Creek
	All taxa	Wise and Molles (1979)	See above	9	Baskets with gravel	River

Table VI, continued

	All taxa	Khalaf and Tachet (1977)	See above	16	Baskets with gravel	Stream
Diversity (H, $\bar{d}$, etc.)	Ephemeroptera	Ulfstrand et al. (1974)	32 (2,4,8,15,32)	15	Trays with gravel	River
	Ephemeroptera	Khalaf and Tachet (1977)	See above	16	Baskets with gravel	Stream
	All taxa (excluding Simuliidae, excluding Simuliidae and Chironomidae)	Khalaf and Tachet (1977)	See above	6	Baskets with gravel	Stream
	Plecoptera	Ulfstrand et al. (1974)	See above	4	Trays with gravel	River
	Trichoptera	Ulfstrand et al. (1974)	See above	4	Trays with gravel	River
	All taxa	Ulfstrand et al. (1974)	See above	4–8	Trays with gravel	River
	All taxa	Roby et al. (1978)	See above	14–28	Baskets with porcelain balls	Creek
Total Numbers	Chironomidae	Bournaud et al. (1978)	21 (1,3,7,14,21)	3	Baskets with rocks	River
	Dugesia	Bournaud et al. (1978)	Same	3	Baskets with rocks	River

	All taxa (excluding *Gammarus*)	Bournaud et al. (1978)	Same	3	Baskets with rocks	River
	All taxa	Cover and Harrel (1978)	112 (weekly for 16 weeks)	49	Multiplates	Canal
Total Number of Taxa, Species	All taxa	Cover and Harrel (1978)	Same	42	Multiplates	Canal
	All taxa	Pearson and Jones (1975)	See [a] above	20	Boxes with gravel	River
			See [c] above	20	Artificial weed	River
	All taxa	Townsend and Hildrew (1976)	12 (3,6,9,12)	6	Trays with gravel	Stream

[a] Artificial substrate on gravel bed.
[b] Artificial substrate on muddy bed.
[c] Artificial weed.
[d] Dry weight.
[e] Wet weight.
[f] Type of weight not specified.

Table VII. Summary of Some Aspects of Experimental Design in Artificial Substrate Colonization Studies Using Containers Filled with Gravel, Rocks or Stones (from Table VI)[a]

Element of Design	Bournaud et al. (1978)	Pearson and Jones (1975)	Townsend and Hildrew (1976)	Ulfstrand et al. (1974)	Shaw and Minshall (1980)	Wise and Molles (1979)	Khalaf and Tachet (1977)
Population Measurement(s) Used	Total numbers	$\bar{x}$ no. box^{-1} $\bar{x}$ no. unit^{-1} [b] no. taxa	$\bar{x}$ no. tray^{-1} no. species	Total numbers diversity (H)	$\bar{x}$ no. 625 cm^{-2} $\bar{x}$ g 625 cm^{-2}	$\bar{x}$ no. basket^{-1} $\bar{x}$ no. species basket^{-1}	$\bar{x}$ no. taxa basket^{-1} $\bar{x}$ no. individuals basket^{-1} Diversity ($\bar{d}$)
Exposure Time (sampling intervals) (days)	21 (1,3,7,14,21)	90 (4,7,14,20, 40,90)	12 (3,6,9,12)	32 (2,4,8,15,32)	64 (1,2,4,8, 16,32,64)	19 (1,2,5,9,19)	28 (2,4,6,8,12,16, 20,24,28)
Placement in Relation to Substrate	On the bottom and suspended in mid-water	On the bottom	On the bottom and suspended 5 cm above the bottom	Implanted	On a plywood platform on the bottom	On the bottom	On the bottom (?)
Gravel, Rock or Stone Sizes Used in Artificial Substrates	Unspecified (although range of total surface areas given)	≥1 cm diameter	Unspecified	Unspecified	phi = −5 (~32–64 mm diam.)	10–25 mm diam., >75 mm diameter and 1:1 mixture of both	16–32 mm diameter
Colonization History of Gravel Fill Used	Unspecified	Unspecified	"from the stream. . .and thoroughly washed to remove all invertebrates" (p. 760)	Previously uncolonized (from a nearby quarry)	From the creek but scrubbed with a stiff brush	Previously uncolonized (bought in Albuquerque)	From the stream but carefully brushed and cleaned

Size of Sampler	Modified waste-paper basket; 25 cm high, 19 cm diameter at base, 28 cm diam. at top	$20 \times 10 \times 5$ cm boxes with open ends	$36.6 \times 17.5 \times 2.5$ cm metal trays	Circular trays, 0.25 $m^2 \times 8$ cm high rim	$25 \times 25 \times 5$ cm wooden trays	$25 \times 25 \times 8$ cm[c] wire baskets	$16 \times 12.5 \times 12.5$ cm plasticized wire cages
Date(s) of Installation	May, June, & July 1976	July 1973	July 1975	April 1973	Series 1 – July 1972 Series 2 – Sept. 1972 Series 3 – Dec. 1972 Series 4 – March 1973	April 1976	July 1974
Sampling Design	Baskets installed at different times and removed after known period of colonization	Unspecified	All trays installed initially and removed sequentially	All trays installed initially and removed sequentially	All trays installed initially, cleaned and reinstalled at preplanned times, and all removed on the same day	All baskets installed initially and removed sequentially	All baskets installed initially and removed sequentially
Retrieval Net Used?	Retrieved into a bucket while underwater	Ends of box closed on withdrawal	Unspecified	Yes	Yes	Unspecified	Unspecified
Mesh Size Used for Sieving	500 μm	Unspecified	Unspecified	Unspecified	Unspecified	Unspecified	Unspecified

[a]All studies were done in running water.
[b]Artificial weed.
[c]Listed in Wise and Molles (1979, p. 70) as $25 \times 25 \times 8$ *mm*, which must be a misprint.

artificial substrates is crucial to making well-informed decisions based on water quality monitoring programs.

Until a better understanding of the operation of artificial substrates develops, we suggest that pilot studies should be done to calibrate use of artificial substrates in particular habitats. Users should know if, and how long, populations take to reach equilibrium levels; how selective the artificial substrate may be; and the influence of seasonal variation in their particular study site. Caution should be exercised in the use of literature values because there is considerable variation between studies in terms of locations, habitats and experimental design, and uncritical and untested recommendations frequently appear in the literature.

Disadvantage 2: Nonrepresentative Sampling Under Either Natural or Polluted Conditions

This is a corollary of the previously discussed statement that artificial substrates are selective (see Disadvantage 1). Therefore, the nonrepresentativeness of artificial substrates will not be discussed in detail in this section and the reader is directed to the above discussion on selectivity.

Pollution Studies. Interpretations of the ability of artificial substrates to measure the effects of pollution differ in the literature. For example, Hughes (1975) casts doubt on their usefulness, whereas Rabeni and Gibbs (1978) and Deutsch (1980) found them to be adequate. Undoubtedly, conclusions about adequacy will depend on the goals of the study (Rabeni and Gibbs 1978).

Use of artificial substrates in studies of pollution should be considered relative (i.e., unknown units allowing comparisons only in space and time) rather than absolute (i.e., density per unit of habitat) measures of macroinvertebrate populations (*sensu* Southwood 1978) because of the difficulty in relating artificial substrate samples to a unit area of stream bottom. In fact, the "relative estimate" feature of artificial substrates applies to most uses of these devices. Relative estimates are also the type obtained with light trap or drift net samples; the biological interpretation of relative estimates is extremely difficult (Southwood 1978).

Placement of Artificial Substrates. Artificial substrates hung in the water column are unrepresentative of bottom conditions (Weber 1973; Rabeni and Gibbs 1978) because different faunas are sampled in these two habitats (Khalaf and Tachet 1978, Figure 1). Benfield et al. (1974) and Bournaud et al. (1978) found qualitative and quantitative differences in taxa collected on bottom artificial substrates compared to ones suspended in the water. Roby et al. (1978) found that burial of samplers by sedimentation resulted in a different fauna than that collected by samplers that remained in the water column and were populated by drifting organisms (see Disadvantage 5). The type of artificial substrate as well as its placement position in the water body also may be important. For example, artificial substrates made of "conservation webbing" appeared to collect higher numbers of organisms in the water column than did concrete cones, but the reverse was true when these two artificial substrates were placed on the bottom (Benfield et al. 1974, Table 1). Bournaud et al. (1978) suggest that macroinvertebrates collected in suspended rather than bottom baskets more closely resembled the natural benthos, although the data of Benfield et al. (1974, Tables 2, 3) indicate quantitative and qualitative differences between artificial substrates located in both positions and kick net samples.

The problem of nonrepresentativeness of suspended artificial substrates can be lessened, in some cases, by placing samplers in appropriate bottom habitats through the use of scuba (e.g., Rabeni and Gibbs 1978; Deutsch 1980; (see also Advantage 2). Placement of artificial substrates in a particular habitat may provide a useful approach for examining features directly affecting that habitat (e.g., Voshell and Simmons 1977; Rabeni and Gibbs 1978).

We conclude that nonrepresentativeness of artificial substrates is a disadvantage as claimed, but one that can be corrected. A more representative sample can be taken if artificial substrates are made to resemble the natural habitat as closely as possible (e.g., RAS) (Minshall and Minshall 1977; Shaw and Minshall 1980) and if they are placed directly into the habitat to be studied.

Disadvantage 3: Artificial Substrates Require Long Exposure Time to Obtain a Sample

Relatively long sampling periods are typical of passive sampling methods (Hellawell 1978) and are a requirement for the use of artificial substrates. The design of sampling programs should take this requirement into account.

Sheldon (1977) examined the possibility of using initial rates of colonization, rather than equilibrium densities, to characterize collections made by artificial substrates. His proposal would minimize the adverse effects (i.e., loss of samplers, successional changes) of leaving artificial substrates in situ for long periods of time. Although preliminary investigations were only partly successful, this approach deserves further attention.

There is little agreement on whether artificial substrates can be used effectively in short-term surveys/pollution monitoring (e.g., see discussion by Lane in Pearson and Jones 1975), but such use for other than qualitative sampling is usually questionable (e.g., Lewis 1980). We conclude that little can be done to lessen this disadvantage since long exposure times are inevitable when using artificial substrates. Users should attempt to visit study sites between designated sampling periods to record events in sampler history, such as water level fluctuations resulting in temporary exposure of the sampler that may be important to interpreting results.

Disadvantage 4: Loss of Fauna on Retrieval of Samplers

Intuitively, one would expect that species lacking adaptations for "clinging," such as Mollusca and Crustacea (Rabeni and Gibbs 1978) and certain species of highly active groups such as Ephemeroptera, would be lost when artificial substrates are retrieved from habitats being sampled. However, studies that have examined this problem report conflicting results. Some authors report negligible losses when natural (Glime and Clemons 1972: *Fontinalis* moss) or artificial (Mason 1977: *Littorella;* Mason et al. 1973: rock-filled baskets and multiplates) substrates are removed from the water. Other authors report that significant losses do occur [Rabeni and Gibbs 1978: rock-filled baskets; Gersabeck and Merritt 1979: ceramic tiles and plastic tapes; Rosenberg and Wiens unpublished data: rock-filled baskets, Table VIII. See also Stanford and Reed (1974) and Pearson and Jones (1975)].

Such losses can be significant sources of variance and also accentuate "nonrepresentativeness" in the use of artificial substrates, but many workers ignore this aspect of sampling (see Chapter 7, Table I). The logical approach is to consider that even careful removal will result in significant losses unless this is checked using a fine mesh net. We would recommend that the sampler be designed to prevent loss of organisms on retrieval (e.g., Mundie 1956; Bull 1968; Hilsenhoff 1969; Crowe 1974; Stanford and Reed 1974; Ulfstrand et al. 1974; Pearson and Jones 1975; Mason 1976; Minshall and Minshall 1977; Kathman 1978; Roby et al. 1978; Armitage 1979; Barber et al. 1979; Page and Neitzel 1979), or that a net of relatively fine mesh be used when the samplers are collected (e.g., Mundie 1956; Macan and Kitching 1972, 1976; Crossman and Cairns 1974; Gale and Thompson 1974; Cover and Harrel 1978; Rabeni and Gibbs 1978; Roby et al. 1978; Gerrish and Bristow 1979; Gersabeck and Merritt 1979; Kovalak 1979; Page and Neitzel 1979). SCUBA can be used to enclose deeply set samplers with fine mesh nets before retrieval (e.g., Gale and Thompson 1974). Macan and Kitching (1972) and Macan

Table VIII. Percentage of Invertebrates Lost from Rock-Filled Basket Artificial Substrates in the Trail and Martin Rivers, Northwest Territories, Canada (Brunskill et al. 1973, Rosenberg and Wiens 1976), during Retrieval (Rosenberg and Wiens 1972)[a]

Taxon	n	Number in Baskets	Number in Dip Net	% of Total Number in Dip Net
Plecoptera	30	351	81	19 (8)
Ephemeroptera	30	1,857	2,891	61 (30)
Trichoptera	30	3,141	605	16 (26)
Chironomidae	30	10,805	2,255	17 (29)
Simuliidae	20	454	50	10[b]
Hydracarina	30	244	100	29
(Mollusca)				(92)
(Crustacea)				(68)

[a]Mesh size of dip net used was 202 μm; n = number of samples. The data of Rabeni and Gibbs (1978), using a 250-μm mesh collection bag, are presented in parentheses for comparison.

[b]Gersabeck and Merrit (1979) also found ≅10% of Simuliid larvae released when the artificial substrates used were disturbed.

(1977) reported that use of a retrieval net for artificial pond weed provided a further advantage by allowing even smaller specimens to be caught than when sweeping was done with a coarser mesh dip net through real vegetation.

Weber (1973) recommended enclosing samplers with at least a 30-mesh ($\sim 600\ \mu m$) dip net on retrieval, although we feel that 200 to 400 μm mesh is preferable. Whatever mesh size is used should be consistent with the one used in processing the samples.

We conclude that loss of fauna on retrieval is definitely a disadvantage that must be considered when using artificial substrates. However, this disadvantage can be minimized by following the recommendations presented above.

Disadvantage 5: Unforeseen Losses of Artificial Substrates

Sampler loss (or altered effectiveness) has hampered many studies in which artificial substrates have been used. Artificial substrates are vulnerable to such extreme water level fluctuations (high water or spate, or low water or drought) (e.g., Boobar and Granett 1978; Fredeen and Spurr 1978; Roby et al. 1978; Wise and Molles 1979; Peckarsky 1980); burial through sedimentation (e.g., Mason et al. 1973; Roby et al. 1978; Deutsch 1980); or vandalism (e.g., Disney 1972; Roby et al. 1978; Meier et al. 1979; Wise and Molles 1979; Peckarsky 1980), which can occur during the long exposure period required to obtain a sample.

With a little bit of "benthic ingenuity" and some luck it may be possible to circumvent losses or altered effectiveness of artificial substrates. Sheldon's (1977) investigation of the possibility that rate of colonization during the first few days after installation (rather than equilibrium densities) could be measured and used to compare colonization in different habitats and by different species is an example of a creative new approach to solving problems in the use of artificial substrates. In addition to encouraging this type of research, we suggest application of the following general strategies to reduce unforseen losses of artificial substrates:

1. *Know the Watershed in Which You Plan to Sample.* Known periods of spate accompanying high rates of transport of sediment likely to bury samplers and low water periods can be avoided (e.g., Mason et al. 1973; Mason 1976; Meier et al. 1979). Likewise, areas heavily frequented by the public should be avoided. A pilot study (e.g., Mason 1976) will help not only to determine sampler efficiences but also to gain familiarity with the study site.

2. *Use Adequate Numbers of Replicates Made of Inexpensive Materials.* Installation of extra artificial substrate replicates can compensate

for losses (e.g., Fredeen and Spurr 1978; Roby et al. 1978; Meier et al. 1979). Cost is a relative matter. For example, Roby et al. (1978), supported by a university grant, were concerned with the loss of samplers costing $8.00 apiece, whereas Wefring and Teed (1980), supported by The Institute of Paper Chemistry, considered a cost of $25.00 per sampler to be inexpensive. Therefore, if artificial substrates are made of inexpensive synthetic materials (e.g., Hellawell and Jellings, in Hellawell 1978; Myers and Southgate 1980), or natural materials such as rocks from the stream bank (Rosenberg and Wiens 1976), small logs (McLachlan 1970; Nilsen and Larimore 1973), mango leaves and palm fronds (Disney 1972), the almost certain loss of some samplers will neither hinder experimental design nor be a financial disaster.

Additional suggestions for minimizing losses or altered effectiveness of artificial substrates are made in Table IX. Luck certainly plays a part in the prevention of sampler losses. Despite the efforts of Roby et al. (1978) to design their samplers adequately and despite placing them in a "restricted, patrolled watershed," (p. 6), losses of samplers produced gaps in their data.

We conclude that artificial substrates are susceptible to unforeseen losses and that this is a disadvantage in their use. However, the effect of unforseen losses on fulfilling study objectives can be minimized.

Disadvantage 6: Inconvenient to Use and Logistically Awkward

Passive samplers generally require two visits to obtain one sample: one to set the sampler and one to recover it (Hellawell 1978). This apparent drawback to the use of artificial substrates (Table II) can be minimized if, in routine surveillance programs, collection of the exposed sampler and setting a new one can be done simultaneously (Hellawell 1978). The initial setting and final retrieval of samplers will usually still require extra visits. Aquatic biologists must decide whether the use of artificial substrates for their individual programs is sufficiently advantageous to offset the extra effort involved.

Often-cited problems in the handling of artificial substrates (Table II) include: (1) the weight and cumbersome nature of some containers filled with organic or inorganic materials; and (2) the difficulties involved in transporting, installing, recovering and processing artificial substrates. Possible solutions to these practical objections are discussed in Table X, but not all problems can be solved. The pros and cons of using different artificial substrate samplers (e.g., Mason et al. 1973; McConville 1975) should be considered before a choice of type is made.

We conclude that the applicability of this disadvantage varies with the

Table IX. Ways to Minimize Unforseen Losses (or altered effectiveness) of Artificial Substrates

Mode of Loss	Suggested Remedies	Comment(s)
Spate, High Water, Fast Current	1. Use scuba.	Samplers can be placed in large deep rivers with high current velocities and sudden large and irregular fluctuations in discharge (e.g., Gale and Thompson 1974; Stanford and Reed 1974; Page and Neitzel 1979; Deutsch 1980).
	2. Use appropriate equipment design/installation:	
	(a) Accommodate water level fluctuations.	Use a float-sampling unit (e.g., Mason et al. 1973; Wefring and Teed 1980).
	(b) Use rugged construction and make samplers heavy	Hilsenhoff (1969) claims his sampler is rugged enough for use in any stream and heavy enough not to be moved by the current. Lewis and Bennett (1974, p. 774) state that the ceramic tiles they used "were sufficiently heavy to maintain position on the stream bed without additional anchorage under all but extreme conditions of spate." The simple design of these tiles probably also helped. See also Hellawell and Jellings (in Hellawell 1978, p. 63) and Deutsch (1980), below.
	(c) Anchor samplers well.	Pearson and Jones (1975) advised that their samplers should be anchored if the stream in which they are to be used is subject to spate. Barber et al. (1979) weighted their artificial seagrass samplers and staked each corner to prevent disturbance by wave action and currents. Holders of the acrylic plates used as artificial substrates by Deutsch (1980) were ballasted with 30 kg of concrete and staked in the channel of the Susquehanna River. Mason (1976), Cover and Harrel (1978), Fredeen and Spurr (1978), Roby et al. (1978), Meier et al. (1979) and Wefring and Teed (1980) also describe how samplers used were anchored (Table X; see also Table I, Chapter 7).

	(d) Use debris guards.	Zillich (1967) and Gale and Thompson (1974) describe devices to protect artificial substrates in running waters from fouling with debris.
Drought, Desiccation, Low Current Velocity	Place in areas deep enough to avoid low water periods.	e.g., Hilsenhoff (1969); Roby et al. (1978); (cf. scuba).
Burial by Sedimentation	Elevate sampler above bottom.	e.g., Roby et al. (1978), although some samplers were still buried. Deutsch (1980) placed samplers (acrylic plates) $\cong$10 cm above the bottom and tilted them 5° from horizontal (downstream edge) to prevent silt deposition. Some samples on the bottom of the plates were not used because sand and gravel filled in under the plates and touched the plate bottom.
Vandalism	1. Use scuba.	Samplers can be installed out of the reach of the public (Voshell and Simmons 1977) and marked inconspicuously, thereby reducing attraction of the public (e.g., Gale and Thompson 1974; Deutsch 1980).
	2. Install evil omens.	Disney (1972, p. 487), in work done in West Cameron, reports that theft of cords holding artificial substrates and suspended from bridges "was eventually prevented by the use of ju-jus consisting principally of semi-mummified monkey heads. . . ."
	3. Install samplers securely.	Gittins (1976), working at Laguna Beach, CA, an area subject to considerable disturbance by waves and people, fastened the base for her artificial substrates to rock or concrete in the littoral area by using a .22-caliber explosive nail driver. The remainder of the artificial substrate device required screwdrivers and a crowbar for removal. Of 12 samplers, 10 lasted for the entire study and two were lost (after 6 and 10 months, respectively) when hit by concrete slabs from a broken pier.
	4. Camouflage or hide samplers.	The artificial substrates used by Gittins (1976) eventually were camouflaged by plant growth. The slate-grey, unglazed, ceramic tiles used by Lewis and Bennett (1974) were similar (color and texture) to the natural rock substrate common in Newfoundland and Labrador streams. These tiles were inconspicuous and not disturbed by the public.

Table IX, continued

Mode of Loss	Suggested Remedies	Comment(s)
		Hilsenhoff (1969) claims that his samplers were rarely disturbed by the public because they are inconspicuous, heavy and obviously of no commercial value. Jacobi (1971) concealed basket artificial substrates from vandals by anchoring the samplers to concrete blocks on the stream bottom. Parsons and Tatum (1974) ran a length of cord from a sampler in shallow, turbid water to a nearby partially submerged structure, thus avoiding use of visible markers. Wefring and Teed (1980) suggest reducing the size of a marker buoy or painting to camouflage it when using their float-sampling unit in areas frequented by the public.
	5. Make samplers heavy and "indestructible." Make them look valueless.	e.g., Hilsenhoff (1969), see above; Hellawell and Jellings (in Hellawell 1978, p. 63). The last of these characteristics is not always a deterrent.
	6. Use signs to explain to the public the purpose and value of the samplers and the study.	This may be useful in situations where the above remedies cannot be used or are proven inadequate (G. A. Lamberti, personal communication, 1981)

Table X. Ways to Minimize Handling Problems of Artificial Substrates

Problem	Possible Solution(s)
Samplers are heavy, cumbersome, and difficult to store and transport.	1. Switch to lighter sampler (e.g., Fullner 1971, from rock-filled baskets to multiplates) or lighten fill (e.g., Bergersen and Galat 1975, from rocks to coniferous tree bark). These authors calibrated the performance of substitutions against the original, a wise and necessary strategy. Wefring and Teed (1980, p. 28) characterized the modified multiplate samplers used in their study as "easily constructed and transported. . . ."
	2. Try to use naturally occurring materials (e.g., rocks from the stream side) as fill rather than carrying heavy materials (e.g., Roby et al. 1978, porcelain balls). The reduction in variance provided by using standardized, constructed materials may not be worth the effort to transport these materials to the sampling site.
	3. Use samplers that can be dismantled for storage and/or transport to save space. Barbeque baskets require large volumes of space, whereas multiplates can be taken apart for storage or transport. (Objectives of the sampling program will determine the type of sampler used, however).
Samplers are difficult to install, anchor.	1. Use scuba (see Table IX).
	2. Attach to natural or manmade objects (e.g., Wolfe and Peterson 1958 – instream or shoreline logs, boulders, etc.; Jacobi 1971 – bridges, pilings, fallen trees; Disney 1972 – bridges; Mason et al. 1973 – water intake tower; Markosova 1974 – rip-rap of a dam).
	3. Use appropriate equipment design (see also Table IX).
	(a) Anchored harness for suspending artificial substrates in the water column of large rivers (e.g., Mason et al. 1973; Crowe 1974; Wefring and Teed 1980; see also Fredeen and Spurr 1978, who describe anchoring schemes for floating artificial substrates).

Table X, continued

Problem	Possible Solution(s)
	(b) Pins or rods, which are part of the sampler and are driven into the substrate (e.g., Cover and Harrel 1978; Roby et al. 1978; Meier et al. 1979).
	(c) Separate pins or stakes used to anchor sampler or to which sampler is attached (e.g., Mason 1976; Fredeen and Spurr 1978; Barber et al. 1979; Benfield et al. 1974—steel stakes cemented into concrete building blocks).
	(d) "Clotheslines"—a bar or line strung between two stakes; samplers hung from bar/line (e.g., Arthur and Horning 1969; Dickson and Cairns 1972; Nilsen and Larimore 1973; Crossman and Cairns 1974; Fredeen and Spurr 1978). *Caution:* May be a navigation hazard when put across streams/rivers! (Even if "clothesline" is put high enough, tether ropes and samplers may be a navigation hazard).
	(e) Add weight to lighter than water artificial substrates to sink them. For example, Rosenberg and Wiens (unpublished data) weighted stick-filled chicken barbeque baskets (see below) with stones wrapped in chicken wire and attached outside of the basket bottom.
	(f) Self-righting sampler, e.g., Hilsenhoff (1969). Because of heavy concrete base, samplers can be dropped into river (depths up to 1.2 m), will sink correctly oriented and remain in place. (See also Table I, Chapter 7).
Samplers are hazardous to navigation.	1. Put samplers on the bottom rather than suspending them in the water column.
	(a) Deep waters—use scuba (See Table IX).
	(b) Shallow waters—e.g., Hilsenhoff (1969) placed samplers far enough below the surface of the water not to interfere with boats. Use snorkel.
	2. Mark samplers conspicuously. *Caution:* Chances of vandalism increase!

Table X, continued

Problem	Possible Solution(s)
Samplers are difficult to find and recover.	1. Use markers. (*Caution:* Must be inconspicuous in areas frequented by the public or expect vandalism to occur), e.g., flags—Mason (1976); buoys, floats—McConville (1975), Fredeen and Spurr (1978), Armitage (1979), Page and Neitzel (1979), Wefring and Teed (1980). Pearson and Jones (1975)—The cork float attached to the string that was used to close the boxes upon retrieval facilitated discovery and recovery from deep water. Macan (1977)—used numbered and lettered posts, originally installed beside the tarn for survey purposes, to identify positions. Cover and Harrel (1978)—strings were attached to eyebolts of multiplates (on a brick base) and led to stakes on the bank of the canal. 2. Maps—e.g., Parsons and Tatum (1974) drew a simple map of the sampling area showing the approximate location of each substrate and using available landmarks. 3. Adopt suitable strategy (e.g., winter sampling). Glime and Clemons (1972) placed artificial substrates in areas of the stream that would remain open in winter so they could avoid chopping ice. Rosenberg and Wiens (1976 and unpublished data) installed tall trees next to artificial substrates in the Trail River before freeze-up. In the Martin River (Rosenberg and Snow 1975; Rosenberg and Wiens unpublished data) large styrofoam floats with bright orange plywood on two sides and suspending barbeque baskets filled with stones were allowed to freeze in. In each river, opposite shores were marked and distances to artificial substrates along the transect were noted for reestablishment in the winter. 4. Use appropriate design of equipment to simplify handling, e.g., Voshell and Simmons (1977) had difficulty handling the large wire

Table X, continued

Problem	Possible Solution(s)
	baskets originally used to sample Lake Anna reservoir. Ultimately, they switched to smaller samplers (see below). (See also Fullner 1971, above).
Samplers require cleaning in the field.	Use appropriate design of equipment, e.g., multiplates can be recovered, put in a container with preservative and mailed to the laboratory for cleaning (e.g., Fullner 1971; Wefring and Teed 1980). Gale and Thompson (1974) returned samplers to the laboratory in retrieval bags. Even if not moistened these bags reduced evaporation so that most organisms survived and could be sorted live in the laboratory.
Samplers that produce large samples require appreciable time to process (i.e., clean, sort, identify organisms).	1. Taking of samples. (a) Use a simple sampler—Lewis and Bennett (1974) cleaned their ceramic 10 x 10 cm tiles in ≤2 min. by scraping with a ruler directly into a container. (b) "Miniaturize" sampler—e.g., Voshell and Simmons (1977) halved sorting time by scaling down their large wire basket samples to small baskets containing ≅ ¼ of the same substrate materials used (leaves, conservation webbing, limestone rocks). Only the miniature rock samples gave unsatisfactory results. e.g., Rosenberg and Wiens (unpublished data), studying colonization of newly flooded shorelines in the Southern Indian Lake reservoir (cf. Rosenberg et al. 1980), originally used barbeque baskets (Anderson and Mason 1968) containing 24, ≅2.5 cm diameter × ≅23 cm long sticks of *Alnus rugosa, Picea mariana* and *Salix* spp. These artificial substrates produced such large samples that they were replaced with individual sticks (≅2.5 cm diam. × ≅50 cm long) linked together end to end and hung vertically in the water (cf. McLachlan 1970). The stick chains were decidedly less time-consuming to handle and process than the basket samples. Field installation, cleaning, and sorting of a stick chain sample took ≅ ¼,

Table X, continued

Problem	Possible Solution(s)
	≅ 1/6, and ≅ 1/10, respectively, the time required for a basket sample although coefficients of variation were lower for the basket samples (≅ 12%, n = 5) than for single sticks from a chain (≅ 25%, n = 5). 2. Treatment of samples. (a) Use faster sorting methods such as flotation, elutriation, etc. (see Rosenberg 1978; Resh 1979, Table 2). (b) Subsample (see Rosenberg 1978). (i) Unsorted samples—e.g., Rosenberg and Wiens (unpublished data). The presence of filamentous algae in samples obtained from baskets of sticks (see above) made samples tedious to sort and precluded the use of flotation or elutriation techniques. Unsorted samples were effectively subsampled by weight. Using χ^2 tests, numbers of macroinvertebtates in subsampled quadrats were not significantly different. (ii) Sorted samples—e.g., Stanford and Reed (1974). Small organisms obtained by elutriation of inorganic substrate and sieving through 351-μm mesh were poured into a plexiglass box, the box was divided into quarters by insertion of a four-way baffle, and the subsample was drained by randomly removing one of the bottom plugs located at each corner.

type of artificial substrate used and how it is handled. In many cases this disadvantage can be minimized (Table X) and study objectives still can be fulfilled.

FINAL CONSIDERATIONS

The design of a sampling program, including the choice of a sampling device, is ultimately a function of the optimal way to fulfill study objectives. From our analysis it is clear that artificial substrates are used for a variety of purposes. Their choice is sometimes an equally effective alternative to direct benthic sampling; at other times (e.g., experimental studies) it is an improvement. And, still at others, it is far less effective (e.g., using SAS for representative benthic surveys). There is no ready-made sampling design available for testing a range of hypotheses, and artificial substrates should be considered as just one of a variety of tools in a "benthic sampling workshop." The final choice of tool will depend on the task to be done.

Our evaluation of the literature on the use of artificial substrates in studies of freshwater benthic macroinvertebrates has revealed that some of the often-claimed advantages regarding their use are incorrect and other purported advantages are inaccurate; however, some of the advantages are correct. Most purported disadvantages of artificial substrate use are, in fact, disadvantages as claimed. However, the majority of these can be corrected or minimized.

Advantages

There is no doubt that artificial substrates are useful for sampling habitats that are difficult to sample by direct methods; that variability due to operator efficiency in taking a sample is reduced; that artificial substrates are convenient to use because they are usually inexpensive and simple to construct; and that sampling by artificial substrate is less destructive than direct methods to a habitat, although not necessarily to the fauna in that habitat. Several advantages are correct but must be qualified. For example, although artificial substrates increase precision in sampling, information on spatial distribution may be sacrificed in the process. Artificial substrates can yield greater "flexibility" in a sampling program, but this depends on the definition of flexibility. The use of nonbiologists to sample with artificial substrates has potential advantages, but certain drawbacks, such as difficulties posed by using complex artificial substrates and the requirement for effective communication, must be considered prior to adopting this strategy. A number of advantages claimed also depend on the type of artificial substrate being used, i.e., standardization of microhabitat features and convenience of use (e.g., ease of processing samples, size and weight of sampler, ease of handling and ease of surface area calculation). Artificial substrates do not help to standardize the choice of macrohabitat to be sampled, contrary to claims in the literature. This point is important and one that deserves more critical attention in the future.

Disadvantages

We have found that several disadvantages claimed in the use of artificial substrates are correct. Many of these disadvantages can be minimized, but this may require considerable effort for some, such as the unknown time for fauna to reach equilibrium levels and sampler selectivity. But it requires less for others, such as losses of fauna on sampler retrieval and unforeseen losses of samplers. However, little can be done presently to alter the requirement for relatively long exposure times, although important events in sampler history can be recorded by visits to the study site between designated sampling periods. Two disadvantages – nonrepresentativeness, inconvenience of use and logistic awkwardness – are correctly claimed, but their severity depends on the type of artificial substrate used and how it is handled. Both can be corrected. Seasonal variation is a factor in all types of sampling; judicious use of artificial substrates can minimize seasonal variations.

Conclusions

Of all the advantages and disadvantages repeatedly claimed for artificial substrate use, we consider the most serious problem to be the incorrect claim that artificial substrates help standardize sampling at the macrohabitat level. The correct use of artificial substrates, as with any other sampling method, requires that they be placed in similar macrohabitats if standardization is to be achieved.

The lack of knowledge of colonization dynamics (selectivity, optimal exposure time) is currently the most serious disadvantage in the use of artificial substrates. Continued experience with artificial substrates will solve many of the problems discussed above, but only a deliberate program of research will solve deficiencies in our knowledge of colonization dynamics. Until this is done, pilot studies to evaluate selectivity and optimal exposure times must be essential components of experimental designs using artificial substrates.

ACKNOWLEDGMENTS

We thank the Freshwater Institute for providing financial support for D. M. Rosenberg. We thank the Office of Water Research and Technology, U.S.D.I., under the allotment program of Public Law 88-379, as amended, and the University of California Water Resources Center, as part of the Office of Water Research and Technology Project No. A-063-CAL and Water Resources Center Project UCAL-WRC-W-519, for providing financial support for V. H. Resh. A. P. Wiens helped at all stages in the production of the paper. S. S. Balling, P. Campbell, I. J. Davies, R. E. Hecky, G. A. Lamberti, E. McElravy, D. F. Malley, K. Patalas, and two anonymous referees reviewed drafts of the manuscript. G. A. Lamberti graciously provided unpublished data. D. Taite drew the figures. D. Laroque typed the manuscript and L. McIver guided it through the word processor.

REFERENCES

Allan, J. D. "The Distributional Ecology and Diversity of Benthic Insects in Cement Creek, Colorado," *Ecology* 56: 1040–1053 (1975).

Allen, K. R. "The Distribution of Stream Bottom Fauna," *N.Z. Ecol. Soc. Proc.* 6: 5–8 (1959).

American Public Health Association (APHA). "Ch. 1005. Benthic Macroinvertebrates," in *Standard Methods for the Examination of Water and Wastewater,* 14th ed. (Washington, DC: American Public Health Association, 1975), pp. 1060–1075.

Anderson, J. B., and W. T. Mason, Jr. "A Comparison of Benthic Macroinvertebrates Collected by Dredge and Basket Sampler," *J. Water Poll. Control Fed.* 40: 252–259 (1968).

Armitage, P. D. "A Folding Artificial Substratum Sampler for Use in Standing Water," *Hydrobiologia* 66: 245–248 (1979).

Arthur, J. W., and W. B. Horning, II. "The Use of Artificial Substrates in Pollution Surveys," *Am. Midl. Nat.* 82: 83–89 (1969).

Barber, W. E., J. G. Greenwood and P. Crocos. "Artificial Seagrass—A New Technique for Sampling the Community," *Hydrobiologia* 65: 135–140 (1979).

Beak, T. W., T. C. Griffing and A. G. Appleby. "Use of Artificial Substrate Samplers to Assess Water Pollution," in *Biological Methods for the Assessment of Water Quality. ASTM Spec. Tech. Publ. 528,* J. Cairns, Jr.

and K. L. Dickson, Eds. (Philadelphia: American Society for Testing and Materials, 1973), pp. 227–241.

Benfield, E. F., A. C. Hendricks and J. Cairns, Jr. "Proficiencies of Two Artificial Substrates in Collecting Stream Macroinvertebrates," *Hydrobiologia* 45: 431–440 (1974).

Bergersen, E. P., and D. L. Galat. "Coniferous Tree Bark: A Lightweight Substitute for Limestone Rock in Barbeque Basket Macroinvertebrate Samplers," *Water Res.* 9: 729–731 (1975).

Boobar, L. R., and J. Granett. "Evaluation of Polyethylene Samplers for Black Fly Larvae (Diptera:Simuliidae), with Particular Reference to Maine Species," *Can. J. Zool.* 56: 2245–2248 (1978).

Bournaud, M., G. Chavanon and H. Tachet. "Structure et fonctionnement des ecosystèmes du Haut-Rhône Français. 5. Colonisation par les macroinvertébrés de substrats artificiels suspendus en pleine eau ou posés sur le fond," *Int. Ver. Theor. Angew. Limnol. Verh.* 20: 1485–1493 (1978).

Brunskill, G. J., D. M. Rosenberg, N. B. Snow, G. L. Vascotto and R. Wagemann. "Ecological Studies of Aquatic Systems in the Mackenzie-Porcupine Drainages in Relation to Proposed Pipeline and Highway Developments," *Can. Task Force North. Oil Dev. Environ. Soc. Comm.* Vol. II, Appendix Rep. 73-41 (1973), pp. 1–345.

Bull, C. J. "A Bottom Fauna Sampler for Use in Stony Streams," *Prog. Fish Cult.* 30: 119–120 (1968).

Cairns, J., Jr., and K. L. Dickson. "A Simple Method for the Biological Assessment of the Effects of Waste Discharges on Aquatic Bottom-Dwelling Organisms," *J. Water Poll. Control Fed.* 43: 755–772 (1971).

Calow, P. "A Method for Determining the Surface Areas of Stones to Enable Quantitative Density Estimates of Littoral Stonedwelling Organisms to Be Made," *Hydrobiologia* 40: 37–50 (1972).

Chutter, F. M., and R. G. Noble. "The Reliability of a Method of Sampling Stream Invertebrates," *Arch. Hydrobiol.* 62: 95–103 (1966).

Cover, E. C., and R. C. Harrel. "Sequences of Colonization, Diversity, Biomass, and Productivity of Macroinvertebrates on Artificial Substrates in a Freshwater Canal," *Hydrobiologia* 59: 81–95 (1978).

Crossman, J. S., and J. Cairns, Jr. "A Comparative Study Between Two Different Artificial Substrate Samplers and Regular Sampling Techniques," *Hydrobiologia* 44: 517–522 (1974).

Crowe, J.—A.M.E. "Tests of Artificial Samplers for Collecting Stream Macroinvertebrates in Manitoba," *Manitoba Entomol.* 8: 19–31 (1974).

Davies, B. R. "The Dispersal of Chironomidae Larvae: A Review," *J. Entomol. Soc. South Afr.* 39: 39–62 (1976).

Deutsch, W. G. "Macroinvertebrate Colonization of Acrylic Plates in a Large River," *Hydrobiologia* 75: 65–72 (1980).

Dickson, K. L., and J. Cairns, Jr. "The Relationship of Fresh-water Macroinvertebrate Communities Collected by Floating Artificial Substrates to the MacArthur-Wilson Equilibrium Model," *Am. Midl. Nat.* 88: 68–75 (1972).

Dickson, K. L., J. Cairns, Jr. and J. C. Arnold. "An Evaluation of the Use of a Basket-Type Artificial Substrate for Sampling Macroinvertebrate Organisms," *Trans. Am. Fish. Soc.* 100: 553–559 (1971).

Disney, R. H. L. "Observations on Sampling Pre-Imaginal Populations of Blackflies (Dipt., Simuliidae) in West Cameroon," *Bull. Entomol. Res.* 61: 485–503 (1972).

Elbert, S. A. "Diversity and Relative Abundance of Macrobenthos in the Ohio River as Determined by Sampling with Artificial Substrates," Ph.D. Thesis, University of Louisville, Louisville, KY (1978).

Elliott, J. M. "Some Methods for the Statistical Analysis of Samples of Benthic Invertebrates. 2nd ed.," *Freshwater Biol. Assoc. Sci. Publ.* 25: 1–160 (1977).

Fredeen, F. J. H., and D. T. Spurr. "Collecting Semi-quantitative Samples of Black Fly Larvae (Diptera:Simuliidae) and Other Aquatic Insects from Large Rivers with the Aid of Artificial Substrates," *Quaest. Entomol.* 14: 411–431 (1978).

Fullner, R. W. "A Comparison of Macroinvertebrates Collected by Basket and Modified Multiple-Plate Samplers," *J. Water Poll. Control Fed.* 43: 494–499 (1971).

Gale, W. F., and J. D. Thompson. "Placement and Retrieval of Artificial Substrate Samplers by SCUBA," *Prog. Fish Cult.* 36: 231–233 (1974).

Gerrish, N., and J. M. Bristow. "Macroinvertebrate Associations with Aquatic Macrophytes and Artificial Substrates," *J. Great Lakes Res.* 5: 69–72 (1979).

Gersabeck, E. F., Jr., and R. W. Merritt. "The Effect of Physical Factors on the Colonization and Relocation Behavior of Immature Black Flies (Diptera:Simuliidae)," *Environ. Entomol.* 8: 34–39 (1979).

Gittins, B. T. "A Littoral Artificial Substrate," *Br. Phycol. J.* 11: 383–386 (1976).

Glime, J. M., and R. M. Clemons. "Species Diversity of Stream Insects on *Fontinalis* spp. Compared to Diversity on Artificial Substrates," *Ecology* 53: 458–464 (1972).

Hales, D. C. "Stream Bottom Sampling as a Research Tool," *Proc. Utah Acad. Sci. Arts Lett.* 39: 84–91 (1962).

Hart, D. D. "Diversity in Stream Insects: Regulation by Rock Size and Microspatial Complexity," *Int. Ver. Theor. Angew. Limnol. Verh.* 20: 1376–1381 (1978).

Hart, D. D., and V. H. Resh. "Movement Patterns and Foraging Ecology of a Stream Caddisfly Larva," *Can. J. Zool.* 58: 1174–1185 (1980).

Hellawell, J. M. "Macroinvertebrate Methods," in *Biological Surveillance of Rivers. A Biological Monitoring Handbook* (Dorchester, England: Dorset Press, 1978), pp. 35–90.

Herrmann, J. "Population Dynamics of *Dendrocoelum lacteum* (O. F. Müller) (Turbellaria, Tricladida) in a South Swedish Lake," *Arch. Hydrobiol.* 85: 482–510 (1979).

Hilsenhoff, W. L. "An Artificial Substrate Device for Sampling Benthic Stream Invertebrates," *Limnol. Oceanog.* 14: 465–471 (1969).

Hornig, C. E., and J. E. Pollard. "Macroinvertebrate Sampling Techniques for Streams in Semi-Arid Regions. Comparison of the Surber Method and a Unit-Effort Traveling Kick Method," U.S. Environmental Protection Agency Rep. 600/4-78-040 (1978), pp. 1–21.

Hughes, B. D. "A Comparison of Four Samplers for Benthic Macroinvertebrates Inhabiting Coarse River Deposits," *Water Res.* 9: 61–69 (1975).

Hynes, H. B. N. *The Ecology of Running Waters.* (Toronto: University of Toronto Press, 1970), pp. 1-555.

Jacobi, G. Z. "A Quantitative Artificial Substrate Sampler for Benthic Macroinvertebrates," *Trans. Am. Fish. Soc.* 100: 136-138 (1971).

Kathman, D. "Artificial Substrate Sampler for Benthic Invertebrates in Ponds, Small Lakes, and Reservoirs," *Prog. Fish. Cult.* 40: 114-115 (1978).

Khalaf, G., and H. Tachet. "La dynamique de colonisation des substrats artificiels par les macroinvertébrés d'un cours d'eau," *Ann. Limnol.* 13: 169-190 (1977).

Khalaf, G., and H. Tachet. "Un problème d'actualité: revue de travaux en matière d'utilisation des substrats artificiels pour l'échantillonnage des macroinvertébrés des eaux courantes," *Bull. Ecol.* 9: 29-38 (1978).

Khalaf, G., and H. Tachet. "Colonization of Artificial Substrata by Macroinvertebrates in a Stream and Variations According to Stone Size," *Freshwater Biol.* 10: 475-482 (1980).

Kovalak, W. P. "Day-Night Changes in Stream Benthos Density in Relation to Current Velocity," *Arch. Hydrobiol.* 87: 1-18 (1979).

Lamberti, G. A. Personal communication (1981).

Lamberti, G. A., and V. H. Resh. "Substrate Relationships, Spatial Distribution Patterns, and Sampling Variability in a Stream Caddisfly Population," *Environ. Entomol.* 8: 561-567 (1979).

Lamberti, G. A., and V. H. Resh. Unpublished data.

Lapchin, L. "Utilisation de substrats artificiels pour l'étude des populations d'invertébrés benthiques. Résultats préliminaires dans un ruisseau à salmonidés de Bretagne," *Ann. Hydrobiol.* 8: 33-44 (1977).

Leglize, L. "Approche méthodologique de la faune benthique de la Moselle: description d'un substrat artificiel composite et résultats préliminaires," *Cah. Lab. Hydrobiol. Montereau* 7: 61-65 (1978).

Lewis, D. J., and G. F. Bennett. "An Artificial Substrate for the Quantitative Comparison of the Densities of Larval Simuliid (Diptera) Populations," *Can. J. Zool.* 52: 773-775 (1974).

Lewis, J. R. "Options and Problems in Environmental Management and Evaluation," *Helgol. Meeresunters.* 33: 452-466 (1980).

Macan, T. T. "The Fauna in the Vegetation of a Moorland Fishpond as Revealed by Different Methods of Collecting," *Hydrobiologia* 55: 3-15 (1977).

Macan, T. T., and A. Kitching. "Some Experiments with Artificial Substrata," *Int. Ver. Theor. Angew. Limnol. Verh.* 18: 213-220 (1972).

Macan, T. T., and A. Kitching. "The Colonization of Squares of Plastic Suspended in Midwater," *Freshwater Biol.* 6: 33-40 (1976).

Markosova, R. "Seasonal Dynamics of the Periphytic Macrofauna in Carp Ponds in South-West Bohemia," *Vestn. Cesk. Spol. Zool.* 38: 251-270 (1974).

Markosova, R. "Development of the Periphytic Community on Artificial Substrates in Fish Ponds," *Int. Rev. Gesamten Hydrobiol.* 64: 811-825 (1979).

Markosova, R. "Effect of Submerged Substrates Quality on the Periphytic Macrofauna in Ponds," *Vestn. Cesk. Spol. Zool.* 44: 46-52 (1980).

Mason, J. C. "Evaluating a Substrate Tray for Sampling the Invertebrate Fauna of Small Streams, with Comment on General Sampling Problems," *Arch. Hydrobiol.* 78: 51-70 (1976).

Mason, W. T., Jr., J. B. Anderson, R. D. Kreis and W. C. Johnson. "Artificial Substrate Sampling, Macroinvertebrates in a Polluted Reach of the Klamath River, Oregon," *J. Water Poll. Control. Fed.* 42: R315-R328 (1970).

Mason, W. T., Jr., C. I. Weber, P. A. Lewis and E. C. Julian. "Factors Affecting the Performance of Basket and Multiplate Macroinvertebrate Samplers," *Freshwater Biol.* 3: 409-436 (1973).

McConville, D. R. "Comparison of Artificial Substrates in Bottom Fauna Studies on a Large River," *J. Minn. Acad. Sci.* 41: 21-24 (1975).

McLachlan, A. J. "Submerged Trees as a Substrate for Benthic Fauna in the Recently Created Lake Kariba (Central Africa)," *J. Appl. Ecol.* 7: 253-266 (1970).

Meier, P. G., D. L. Penrose and L. Polak. "The Rate of Colonization by Macroinvertebrates on Artificial Substrate Samplers," *Freshwater Biol.* 9: 381-392 (1979).

Merritt, R. W., K. W. Cummins and J. R. Barnes. "Demonstration of Stream Watershed Community Processes with Some Simple Bioassay Techniques," in *Innovative Teaching in Aquatic Entomology,* V. H. Resh and D. M. Rosenberg, Eds. *Can. Spec. Publ. Fish. Aquat. Sci.* 43: 101-113 (1979).

Merritt, R. W., K. W. Cummins and V. H. Resh. "Collecting, Sampling, and Rearing Methods for Aquatic Insects," in *An Introduction to the Aquatic Insects of North America,* R. W. Merritt and K. W. Cummins, Eds. (Dubuque, IA: Kendall-Hunt Publishing Co., 1978), pp. 13-28.

Minshall, G. W., and J. N. Minshall. "Microdistribution of Benthic Invertebrates in a Rocky Mountain (U.S.A.) Stream," *Hydrobiologia* 55: 231-249 (1977).

Moon, H. P. "Methods and Apparatus Suitable for an Investigation of the Littoral Region of Oligotrophic Lakes," *Int. Rev. Ges. Hydrobiol. Hydrog.* 32: 319-333 (1935a).

Moon, H. P. "Flood Movements of the Littoral Fauna of Windermere," *J. Animal Ecol.* 4: 216-228 (1935b).

Moon, H. P. "An Investigation of the Movements of Freshwater Invertebrate Faunas," *J. Anim. Ecol.* 9: 76-83 (1940).

Mundie, J. H. "A Bottom Sampler for Inclined Rock Surfaces in Lakes," *J. Anim. Ecol.* 25: 429-432 (1956).

Myers, A. A., and T. Southgate. "Artificial Substrates as a Means of Monitoring Rocky Shore Cryptofauna," *J. Mar. Biol. Assoc. U.K.* 60: 963-975 (1980).

Needham, P. R., and R. L. Usinger. "Variability in the Macrofauna of a Single Riffle in Prosser Creek, California as Indicated by the Surber Sampler," *Hilgardia* 24: 383-409 (1956).

Nilsen, H. C., and R. W. Larimore. "Establishment of Invertebrate Communities on Log Substrates in the Kaskaskia River, Illinois," *Ecology* 54: 366-374 (1973).

Page, T. L., and D. A. Neitzel. "A Device to Hold and Identify Rock-Filled Baskets for Benthic Sampling in Large Rivers," *Limnol. Oceanog.* 24: 988-990 (1979).

Parsons, D. S., and J. W. Tatum. "A New Shallow Water Multiple-Plate Sampler," *Prog. Fish. Cult.* 36: 179–180 (1974).

Pearson, R. G., and N. V. Jones. "The Colonization of Artificial Substrata by Stream Macro-invertebrates," *Prog. Water Technol.* 7: 497–504 (1975).

Peckarsky, B. L. "Biological Interactions as Determinants of Distributions of Benthic Invertebrates Within the Substrate of Stony Streams," *Limnol. Oceanog.* 24: 59–68 (1979).

Peckarsky, B. L. "Influence of Detritus Upon Colonization of Stream Invertebrates," *Can. J. Fish. Aquat. Sci.* 37: 957–963 (1980).

Pielou, E. C. "The Measurement of Diversity in Different Types of Biological Collections," *J. Theor. Biol.* 13: 131–144 (1966).

Rabeni, C. F., and K. E. Gibbs. "Comparison of Two Methods Used by Divers for Sampling Benthic Invertebrates in Deep Rivers," *J. Fish. Res. Board Can.* 35: 332–336 (1978).

Rabeni, C. F., and G. W. Minshall. "Factors Affecting Microdistribution of Stream Benthic Insects," *Oikos* 29: 33–43 (1977).

Reice, S. R. "The Role of Substratum in Benthic Macroinvertebrate Microdistribution and Litter Decomposition in a Woodland Stream," *Ecology* 61: 580–590 (1980).

Resh, V. H. "Sampling Variability and Life History Features: Basic Considerations in the Design of Aquatic Insect Studies," *J. Fish. Res. Board Can.* 36: 290–311 (1979).

Resh, V. H. Unpublished data.

Roby, K. B., J. D. Newbold and D. C. Erman. "Effectiveness of an Artificial Substrate for Sampling Macroinvertebrates in Small Streams," *Freshwater Biol.* 8: 1–8 (1978).

Rosenberg, D. M. "Practical Sampling of Freshwater Macrozoobenthos: A Bibliography of Useful Texts, Reviews, and Recent Papers," *Can. Fish. Mar. Serv. Tech. Rep.* 790: 1–15 (1978).

Rosenberg, D. M., and N. B. Snow. "Ecological Studies of Aquatic Organisms in the Mackenzie and Porcupine River Drainages in Relation to Sedimentation," *Can. Fish. Mar. Serv. Res. Dev. Tech. Rep.* 547: 1–86 (1975).

Rosenberg, D. M., and A. P. Wiens. "Community and Species Responses of Chironomidae (Diptera) to Contamination of Fresh Waters by Crude Oil and Petroleum Products, with Special Reference to the Trail River, Northwest Territories," *J. Fish. Res. Bd. Can.* 33: 1955–1963 (1976).

Rosenberg, D. M., and A. P. Wiens. Unpublished data.

Rosenberg, D. M., A. P. Wiens and B. Bilyj. "Sampling Emerging Chironomidae (Diptera) With Submerged Funnel Traps in a New Northern Canadian Reservoir, Southern Indian Lake, Manitoba," *Can. J. Fish. Aquat. Sci.* 37: 927–936 (1980).

Shaw, D. W., and G. W. Minshall. "Colonization of an Introduced Substrate by Stream Macroinvertebrates," *Oikos* 34: 259–271 (1980).

Sheldon, A. L. "Colonization Curves: Application to Stream Insects on Semi-Natural Substrates," *Oikos* 28: 256–261 (1977).

Shelford, V. E., and S. Eddy. "Methods for the Study of Stream Communities," *Ecology* 10: 382–391 (1929).

Simberloff, D. "Constraints on Community Structure During Colonization," in *Environmental Biomonitoring, Assessment, Prediction, and Management—Certain Case Studies and Related Quantitative Issues,* J. Cairns, Jr., G. P. Patil and W. E. Waters, Eds. (Fairland, MD: International Co-operative Publishing House, 1979), pp. 415–424.

Slack, K. V., R. C. Averett, P. E. Greeson and R. G. Lipscomb. "Methods for Collection and Analysis of Aquatic Biological and Microbiological Samples," in *Techniques of Water-Resources Investigations of the United States Geological Survey, Book 5* (Washington, DC: U. S. Department of the Interior, Geological Survey, 1973), pp. 1–165.

Sladeckova, A., and E. Pieczynska. "Periphyton," in *A Manual on Methods for the Assessment of Secondary Productivity in Fresh Waters. IBP Handb. 17,* W. T. Edmondson and G. G. Winberg, Eds. (Oxford, England: Blackwell Scientific Publications, 1971), pp. 109–122.

Southwood, T. R. E. *Ecological Methods with Particular Reference to the Study of Insect Populations,* 2nd ed. (London, England: Chapman and Hall, 1978), pp. 1–524.

Stanford, J. A., and E. B. Reed. "A Basket Sampling Technique for Quantifying Riverine Macrobenthos," *Water Resources Bull.* 10: 470–477 (1974).

Townsend, C. R., and A. G. Hildrew. "Field Experiments on the Drifting, Colonization and Continuous Redistribution of Stream Benthos," *J. Animal Ecol.* 45: 759–772 (1976).

Tsui, P. T. P., and B. W. Breedlove. "Use of the Multiple-Plate Sampler in Biological Monitoring of the Aquatic Environment," *Fla. Sci.* 41: 110–116 (1978).

Ulfstrand, S., L. M. Nilsson and A. Stergar. "Composition and Diversity of Benthic Species Collectives Colonizing Implanted Substrates in a South Swedish Stream," *Entomol. Scand.* 5: 115–122 (1974).

Voshell, J. R., Jr., and G. M. Simmons, Jr. "An Evaluation of Artificial Substrates for Sampling Macrobenthos in Reservoirs," *Hydrobiologia* 53: 257–269 (1977).

Wanson, M., and C. Henrard. "Habitat et comportement larvaire du *Simulium damnosum* Theobald," *Recl. Trav. Sci. Med. Congo Belge* 4: 113–121 (1945).

Waters, W. E., and V. H. Resh. "Ecological and Statistical Features of Sampling Insect Populations in Forest and Aquatic Environments," in *Contemporary Quantitative Ecology and Related Econometrics,* G. P. Patil and M. Rosenzweig, Eds. (Fairland, MD: International Co-operative Publishing House, 1979), pp. 569–617.

Weber, C. I., Ed. "Biological Field and Laboratory Methods for Measuring the Quality of Surface Waters and Effluents," *U.S. Environmental Protection Agency, Environmental Monitoring Series,* EPA-670/4-73-001 (1973), pp. 1–186.

Wefring, D. R., and J. C. Teed. "Device for Collecting Replicate Artificial Substrate Samples of Benthic Invertebrates in Large Rivers," *Prog. Fish Cult.* 42: 26–28 (1980).

Wene, G., and E. L. Wickliff. "Modification of a Stream Bottom and Its Effect on the Insect Fauna," *Can. Entomol.* 72: 131–135 (1940).

Williams, D. D. "Movements of Benthos During the Recolonization of Temporary Streams," *Oikos* 29: 306–312 (1977).

Williams, D. D. "Temporal Patterns in Recolonization of Stream Benthos," *Arch. Hydrobiol.* 90: 56–74 (1980).

Williams, D. D., and H. B. N. Hynes. "The Recolonization Mechanisms of Stream Benthos," *Oikos* 27: 265–272 (1976a).

Williams D. D., and H. B. N. Hynes. "Stream Habitat Selection by Aerially Colonizing Invertebrates," *Can. J. Zool.* 54: 685–693 (1976b).

Wise, D. H., and M. C. Molles, Jr. "Colonization of Artificial Substrates by Stream Insects: Influence of Substrate Size and Diversity," *Hydrobiologia* 65: 69–74 (1979).

Wolfe, L. S., and D. G. Peterson. "A New Method to Estimate Levels of Infestations of Black Fly Larvae (Diptera:Simuliidae)," *Can. J. Zool.* 36: 863–867 (1958).

Zillich, J. A. "Responses of Lotic Insects to Artificial Substrate Samplers," M.Sc. Thesis, University of Wisconsin (1967).

CHAPTER 7

TYPES OF ARTIFICIAL SUBSTRATES USED FOR SAMPLING FRESHWATER BENTHIC MACROINVERTEBRATES

John F. Flannagan and David M. Rosenberg

Department of Fisheries and Oceans
Freshwater Institute
Winnipeg, Manitoba R3T 2N6

Our objective is to provide an overview of the literature on artificial samplers that have been used for sampling freshwater benthic macroinvertebrates. There is a need in the literature for a complete and annotated bibliography of artificial substrate sampler types. Such a list is presented in Table I. Studies comparing the performance of artificial substrates to other artificial substrates and to direct sampling methods (e.g., Hess, Surber, Kick, grab, etc.), are summarized in Table II.

Table I identifies eight major categories of artificial substrates:

1. containers filled with various substrates
2. multiplates
3. boards, panels and tiles
4. bricks and blocks
5. plastic sheets, polyethylene and fabric strips, ropes, etc.
6. implanted substrates
7. natural organic substrates
8. miscellaneous substrates

Types of equipment within each category are described and arranged chronologically (oldest to most recent) according to our judgment of increased complexity.

Table I. Kinds of Artificial Substrate Samplers Described in the Literature

Description	Reference	Comment
1. CONTAINERS (see also implanted substrates)		
a) *Rock-filled*		
i) *Pans and trays*		
Plastic trays filled with granite stones of various sizes, suspended below a 300-μm mesh screen.	Laville (1974)	Screen slid down suspending ropes to cover sampler before retrieval.
Circular metal trays (2500 cm^2 × 8 cm deep) filled with stones.	Friberg et al. (1977)	Net used to enclose samples on retrieval.
Trays (37 × 25 × 5 cm) with wooden sides and mesh bottoms, filled with gravel or coarser substrate from stream but without insects or algal film.	Sheldon (1977)	Trays recovered by sliding a metal pan underneath sampler and lifting sampler out.
Trays with wood frame (625 cm^2 × 9.5 cm or 5 cm deep) and perforated hardwood base covered with nylon mesh; filled with stones.	Minshall and Minshall (1977) Shaw and Minshall (1980)	Hand net used to contain samples on retrieval.
ii) *Boxes and baskets*		
Wire baskets (929 cm^2 × 10.2 cm high), 0.64-cm mesh, filled with two sizes of rocks.	Wene and Wickliff (1940)	Used in shallow water. Lifted into a dip net for retrieval.
"Brush boxes"—0.64-cm mesh box filled with rubble.[a]	Scott (1958)	
Barbeque basket (17.8 cm diameter × 28 cm long) filled with limestone chips 2.5–5 cm-diameter; weight ≅10 kg. Attached to a float. Samples taken from bottom or suspended in mid-water.	Mason et al. (1967) Anderson and Mason (1968) Mason et al. (1970) Dickson et al. (1971) Crandall (1977) Rabeni and Gibbs (1978) Fredeen and Spurr (1978)	No method described to prevent losses of invertebrates on retrieval except for Rabeni and Gibbs (1978), who used divers with fine-mesh nylon nets.
Baskets (7.6 or 15 × 15 × 30 cm) of #18 gauge, galvanized hardware cloth,	Zillich (1967)	Retrieved with a dip net. Author compared a

Table I, continued

Description	Reference	Comment
reinforced with wire, filled with limestone chips, 7.6 cm-diameter anchored with poles.		number of different baskets and investigated animal loss if a method to contain samples on retrieval is not used.
Collapsible basket (32 cm diameter × 23 cm high) composed of collar of stiff wire mesh with a fine mesh bag attached below. Bag is filled with gravel and lowered to the bottom where it collapses leaving gravel enclosed by the collar.	Bull (1968)	Usable in deeper waters rather than above samplers. Fine-mesh bag encloses sample on retrieval.
Cylindrical sampler (12.7 cm diameter × 12.7 high) made of solid galvanized iron with 3.6 kg of limestone sandwiched between pieces of 16-gauge galvanized hardware cloth inserted inside clyinder. Apparatus is attached to a large concrete block.	Hilsenhoff (1969)	Also used by Minshall and Andrews (1973). A device to retrieve samplers described.
Apparatus described by simultaneously using barbeque baskets suspended at three depths.[a]	Kreis and Smith (1970) Kreis et al. (1971)	
Open-ended aluminum box with Perspex roof, partly filled with substrate.	Pearson and Jones (1975)	Doors on open ends closed to contain sample on recovery. Wrapped in a cloth when retrieved from shallow water.
Plastic wastepaper basket reinforced with metal rods and having a lid of metal mesh, filled with limestone rocks.	Roux et al. (1976) Bournaud et al. (1978)	Retrieved into a bucket from deeper water. Retrieval into a net would appear to be better.

Table I, continued

Description	Reference	Comment
Baskets (25 × 25 × 8 cm) of 13-mm mesh galvanized screen, filled with small, large or mixed gravel.[a]	Wise and Molles (1979)	
Plastic cages filled with stones of various sizes.[a]	Khalaf and Tachet (1977, 1980)	
Barbeque basket filled with limestone chips, attached vertically with central iron rod to concrete block.	Flannagan et al. (1980)	Designed to work in deep, fast rivers. Equipped with enclosure net.
b) *Other fill*		
i) *Pans trays*		
Canvas bag lashed onto metal frame (60 × 60 cm) and filled with rubble.[a]	Moon (1935)	
Square iron frames with wire netting, filled with defaunated natural substrate.[a]	Moon (1940)	
Aluminum cake pans in pairs, one on each end of an iron bar, and filled with clean mud.[a]	Moore and Cook (1967)	
Iron tray with two discs of expanded metal mesh inside.	Beak et al. (1973)	Heavy enough not to require anchoring except in very fast waters. Equipped with movable metal cover, which is inverted over pan before retrieval.
Square steel trays (40 × 40 × 31 cm deep) with two layers of expanded metal screen, equipped with concrete weighting device.	Crowe (1974)	Sampler enclosed in hand net before retrieval. Tray samplers not as good as other artificial substrates tested.
Alternating arrangement of four trays and four plates made	Goddard et al. (1975)	Organisms observed by

Table I, continued

Description	Reference	Comment
of plywood. Rock surfaces mimicked by painting with sandy vinyl paint.[a]		scuba diver.
Metal trays filled with defaunated stream substrate, either placed on bottom or suspended above it.[a]	Townsend and Hildrew (1976)	Anchoring method not described.
Large (3.1 × 3.7 × 0.15 meter) wooden boxes filled with defaunated mud.	McCall (1977)	Boxes subsampled by divers or using Van Veen grab.
ii) *Boxes and baskets*		
Shallow canvas bag on metal frame filled with *Littorella*.[a]	Moon (1935)	
Baskets filled with *Carex stricta*.	Zillich (1967)	A device to retrieve samplers described.
Barbeque baskets filled with styrofoam, concrete or wood spheres (all 7.5 cm in diameter).[a]	Jacobi (1971)	
Baskets with 10, 7.5-cm-diameter styrofoam balls.	Crowe (1972b)	This sampler recommended by the author over rock-filled baskets. Floating and anchoring mechanism described.
Barbeque baskets filled with three layers of cinder blocks, conservation webbing on top of each layer.	Dhalberg and Conyers (1974)	Samples wrapped in a layer of muslin cloth as removed from the river.
Baskets with concrete cones. Steel rods driven into stream bed used as anchoring devices.[a]	Benfield et al. (1974)	
Barbeque baskets filled with 5 to 7.5-cm pieces of heat-treated commercially available coniferous tree bark.	Bergersen and Galat (1975)	Recommended by authors because of lightness. Samplers collected using a dip net.

Table I, continued

Description	Reference	Comment
Conservation webbing woven into bottom of baskets[a]	Witham et al. (1968)[b] Simmons and Winfield (1971)	
Barbeque baskets with twenty, 10 × 10 cm pieces of conservation webbing[a]	Stauffer et al. (1976) (see also Benfield et al. 1974)	
Conservation webbing (four, 15 cm^2 pieces), doubled, put inside barbeque baskets[a]	Hocutt et al. (1976) Prins and Black (1971)	
Mesh-covered wood-frame boxes containing metal mesh and plant detritus. Box put into mesh basket suspended in river.[a]	Lapchin (1977)	
Baskets containing leaves, rocks or conservation webbing. Perforated plastic cottage cheese containers with the same fill.	Voshell and Simmons (1977)	Samplers collected by scuba divers using cloth flour sacks.
Basket mounted vertically on wood platform fitted with polyester fabric enclosure sack. Basket filled with a variety of artificial substrates.	Kathman (1978)	Designed for use in ponds, lakes, etc. No comparative data given.
Baskets with 25 porcelain balls (5 cm in diameter), layers of 0.3-mm screening under the lower layer of balls.	Roby et al. (1978)	Sampler retrieved with aid of dip net. Authors suggest that carefully taken Surbers are just as good and less of a problem than these samplers.
Brass framework supporting a square of plastic webbing covered with a folding net bag.	Armitage (1979)	Has possibilities for use in deep water.
2. MULTIPLATES		
Alternating 7.6-cm and 2.5-cm squares of 0.3-cm-thick masonite	Hester and Dendy (1962)	Enclosed in plastic bags

Table I, continued

Description	Reference	Comment
mounted on a centrally positioned bolt.		before removal from water to help minimize loss of animals, but only possible in shallow water.
Modification of Hester-Dendy consisting of two plates separated by a spacer.[a]	Ebert (1965)[b] Pardue (1973)	
Used five sets of Hester-Dendy samplers mounted on a 1.2 meter cross-piece.[a]	Arthur and Horning (1969)	
Hester-Dendy with heavy metal baseplate to allow use in streams.	Mathers and Martin (1969)	Sample retrieved with help of empty coffee can to contain animals.
Hester-Dendy sampler with 14 circular plates and 24 spacers. Suspended or anchored on a rod.[a]	Fullner (1971) Weber (1973) Tsui and Breedlove (1978)	
Round Hester-Dendy with varying distances between plates and with a three-hole brick base.	Parsons and Tatum (1974)	Containing device for recovering sample has cheesecloth screen to allow free passage of water. Sampler can be used with the containing device only in shallow water.
Masonite plates (10 × 10 × 0.32 cm) with ventilation holes, mounted alternately with 10-cm^2 sections of conservation webbing.[a]	McDaniel (1974)	
Parsons and Tatum (1974) modification of Hester-Dendy with plate surfaces alternately smooth and roughened.	Cover and Harrel (1978)	Appears to be ultimate development of Hester-Dendy at least for use in shallow streams. Perhaps similar to Hester-

Table I, continued

Description	Reference	Comment
		Dendy as masonite plates are smooth only on one side.
Hester-Dendy samplers with rough and smooth sides variously arranged.	Harrold (1978)	Author suggests that Hester-Dendy samplers be set up so that the eight plates have rough sides facing each other.
Reduced numbers of plates from Hester-Dendy.	Meier et al. (1979)	Fastened to the bottom by pounding extra rod section into the stream bed.
Twin boards of pine or masonite with space between.[a]	Pardue and Nielsen (1979)	
A floating device for taking triplicate multiplate samples.	Wefring and Teed (1980)	Fine dip net recommended for retrieval, but actual method not described.
3. BOARDS, PANELS AND TILES		
White glazed tiles sunk in stream bed.[c]	Zahar (1951) Hall and Edwards (1978)	Found to be unsatisfactory by author, but more frequent sampling may have helped.
Single circular asbestos-cement plate (41 cm in diameter × 3 mm thick).	Mundie (1956)	Used on bedrock on lake bottoms. With covering cone and bucket to retrieve sampler without loss of substrate or animals.
White polyethylene plate (342 cm^2) attached vertically to a length of plastic pipe. Pipe fitted over another of narrower diameter standing vertically in stream bed. Plate floats just below the surface of the water.[a]	Besch et al. (1967)	

Table I, continued

Description	Reference	Comment
Wooden and black asbestos panels suspended vertically. Device includes float and anchor.[a]	Cory (1967)	
Hardboard strips painted white and ruled in 2.5-cm squares.[c]	Curtis (1968)	Fitted with an angle iron anchor. Author suggests that blackfly larvae detach when artificial substrates are removed. Suggests a photographic method for counting.
Horizontal and vertical devices designed to hold asbestos panels.	Bourget and Lacroix (1971)	Anchoring device and marker buoy also described. Samples collected using scuba divers.
Unglazed ceramic tile laid on stream bed.[a,c]	Lewis and Bennett (1974) Boobar and Granett (1978)	Substrate not disturbed by public because of natural color.
Glass microscope slides held in place by acrylic plates mounted on wood backing boards.	Gittins (1976)	No results presented.
Wooden planks placed vertically and horizontally in lakes.	Konstantinov (1977)	Horizontally placed artificial substrates collected more species and specimens.
Glass plates suspended at various depths.[a]	Protasov (1979)	
White ceramic tiles and clear plastic tapes set at various depths and on bottom[c]	Gersabeck and Merritt (1979)	Sampler enclosed in hand net before removal.
Various devices including plastic sheets, bricks, tiles, etc.[c]	Nakamura et al. (1978)	
Wooden plates suspended from anchored buoys[a]	Markosova (1980)	

Table I, continued

Description	Reference	Comment
Plastic paddles and foil suspended from anchored floats[a]	Markosova (1979, 1980)	
Frosted (by sandblasting) 13-mm-thick acrylic plates (20 × 30 cm) in specially designed, weighted holder, held 10 cm above river bottom.	Deutsch (1980)	Sampled using scuba and sample enclosing device.
4. BRICKS AND BLOCKS		
Concrete blocks 1 (m^2) with grooves.[a]	Shelford and Eddy (1929)	
Concrete blocks (32 × 32 cm) with ring for buoy. Bottom of block with grooves[a]	Britt (1955)	
Concrete and wood blocks in pairs.[a]	Hoar and Miller (1972)	Anchoring device described.
Granite cubes (10 × 10 × 10 cm) suspended at various depths.[a]	Markosova (1974, 1980)	Anchoring and suspending device described.
Sand-case bricks (20.6 × 9.8 × 5.9 cm).	Kovalak (1976)	
As above, but bricks sunk 1 cm into stream bed.	Kovalak (1978, 1979)	Bricks put in hand net for retrieval.
Concrete bricks (20 × 40 cm).	Herrmann (1979)	Used only for flatworms.
Various devices, including bricks and blocks.[c]	Nakamura et al. (1978)	
5. PLASTIC SHEETS, POLYETHYLENE AND FABRIC STRIPS, ROPES, ETC.		
Hollow galvanized metal cones painted white. Held in stream by lengths of wire attached to apexes of cones and fastened to instream or shoreline logs, boulders, etc. Also used white plates for preliminary experiments.[a,c]	Wolfe and Peterson (1958)	
A length of polyethylene tape (50 cm × 2.5 cm) attached to a stone by string.[a,c]	Williams and Obeng 1962) Obeng (1967) Doby et al. (1967)	
Rectangles of fabric of various kinds anchored at one end with pieces of brick.[a,c]	Tarshis (1968)	

Table I, continued

Description	Reference	Comment
Rows of plastic ribbons suspended from a rope stretched across the river.[a,c]	Doby et al. (1967) Elsen and Hebrard (1977)	
Square panels of conservation webbing (10 × 10 cm) attached to a styrofoam float suspended between two posts driven into river bed.[a]	Dickson and Cairns (1972)	
Polypropylene rope (13 mm circumference) unravelled or used as is, woven into a lattice.	Macan and Kitching (1972) Macan (1977)	To simulate vegetation. Retrieved with a large net.
Artificial mosses made from #60 cotton twine (20 × 32-cm lengths) tied in the middle; or from two heavy plastic bags cut to the base (1-cm strips), sewed together and rolled. Both types were attached to the substrate with string.[a]	Glime and Clemons (1972)	
Artificial weeds similar to those of Macan and Kitching (1972).	Pearson and Jones (1975)	Sample not enclosed until brought to water surface.
Artificial plants made from five leaves (each 16.5 cm^2) of polyvinyl chloride (PVC) mounted along an anchored and buoyed string.	Soska (1975)	No retrieval system described.
Flat or pleated sheets (31.6 × 31.6 cm) of plastic suspended in mid-water.	Macan and Kitching (1976)	Retrieved by a net from below.
Compared various artificial substrates and the effects of position, duration of exposure, shape, etc., on density of invertebrates collected.[c]	Pegle and Rühm (1976)	Authors concluded narrow strips better than wide ones.
Strips of plastic attached to a central pole anchored in a concrete block.[c]	Elsen (1977)	
Rectangular sheet of plastic mounted vertically in a metal frame, anchored as above.[c]		
A large stone tied to a depth-graduated rope. This method used to obtain larvae in the deepest parts of the river.[a]		

Table I, continued

Description	Reference	Comment
Lengths of polyethylene tape on 2-meter metal rods. Rods held up vertically in stream bed by rocks.[a]	Boobar and Granett (1978)	
Artificial plants made from stiff white plastic and black electrical tape attached to an anchor and buoyed central string.	Gerrish and Bristow (1979)	Retrieved by a diver by enclosing whole "plant" in a plastic bag before moving it.
Artificial seagrass constructed of polyweave ribbons mounted on pieces 50 × 50 cm 1-mm mesh nylon netting.	Barber et al. (1979)	
Polyethylene strip (1.25 cm wide × 40 cm long) attached to a short wooden stake driven into the substrate.	Pegel (1980)	Retrieved using a dip net.
6. IMPLANTED SUBSTRATES		
a) *Trays*		
Angle iron frames delineating six equal areas, inserted into stream bed. Six masonite boxes bolted onto frame. Samples taken by blocking front of a box with a net and fitting a drift net to the back of each box, then agitating delineated area.	Waters (1964)	Uses natural substrate in natural situation. Questionable if artificial substrate method.
White polystyrene trays (30 × 24 × 5 cm deep) each with six small bricks laid on 10 or 50 ml of plant detritus or without underlying detritus. Tray set into the stream bed for 22 days. Trays also used with rubber fragments instead of plant detritus.	Egglishaw (1964)	No retrieval system described. Experiments showed importance of plant detritus as regulator of animal distribution.
Galvanized mesh trays (20 × 25 × 10 cm) buried in 10-cm-deep holes in stream bottom. Trays filled with the excavated substrate.[a]	Hughes (1975)	
Circular trays (0.25 m^2 × 8 cm deep) with gravel and stones inside nylon mesh bag. Bags rolled down and bags and trays implanted in the substrate.	Ulfstrand et al. (1974) Nilsson and Sjöström (1977)	To remove, bag unrolled and tray enclosed in it.

Table I, continued

Description	Reference	Comment
Steel collar welded to steel mesh base and with metal lid. Made to fit snugly in a Hess sampler. Top 10 cm of stream substrate removed and replaced after sampler inserted. Sampler enclosed in a Hess sampler and sample taken.	Mason (1976)	Nitex liner prevents loss of animals on retrieval. Tray provided with anchoring pin.
b) *Pots, baskets, etc.*		
Plastic buckets with mesh covers buried in substrate.	Schwoerbel (1967) An der Lan (1967)	Designed to sample hyporheos animals.
Pot with flanges buried in the substrate, filled with excavated undisturbed substrate. Top of pot left exposed above substrate. Sample collected by attaching long cylinder to pot flanges, thereby forming a watertight seal and withdrawing whole sampler.	Friedrich and Herbst (1968)	For use in littoral zone. Retrieval cylinder must be longer than depth of water from which sample is taken
Perforated (0.6 mm) juice cans (1.4 liters) buried and filled with excavated substrate[a]	Radford and Hartland-Rowe (1971)	
Three-piece sampler consisting of outer pipe, inner pot (both of aluminum) and folded nylon bag between them. Pipe buried in substrate and sampling pot enclosed in net on retrieval.	Coleman and Hynes (1970)	Authors discuss possible problems with this device as hyporheos sampler.
Metal box with two sliding doors. Stream bed excavated, box implanted with doors open.	Bishop (1973)	Before retrieval, doors closed and whole sample enclosed in plastic bag before lifting from water. Author discusses factors contributing to errors in the sampling method.
Perforated plastic pipe divided horizontally into several compartments and within a perforated steel pipe. Whole apparatus filled with	Hynes (1974)	Described by author as "heavy, awkward,

Table I, continued

Description	Reference	Comment
gravel and pounded into stream bed. Pipe constructed so that perforations in inner and outer pipes can be matched to allow colonization or closed for retrieval.		frustrating... obsolete."
Basket sampler similar to that of Wene and Wickliff (1940) but with concrete base and Nitex® containing net. Sampler partially buried in substrate.	Stanford and Reed (1974)	Enclosed net buried and substrate around sampler must be removed to expose it.
Galvanized metal pipe (30 cm long × 12.5 cm diameter) with screen on bottom end and filled with ten 7.5-cm styrofoam balls.	Crowe (1974)	Considered by author to be inferior to barbeque basket with styrofoam balls.
Square boxes with mesh base filled with stones and gravel from stream bed and buried in stream bed.[a]	Legier and Talin (1975)	
Horizontal plastic pipe (1 or 2 meters long × 12 cm diameter). Bottom covered with substrate, back end with net.	Nilsson and Sjöström (1977)	No anchoring device described. Pipes with leaves and stones colonized more quickly than those with just stones.
Stainless steel mesh and frame boxes, buried in stream bed and filled with natural or other substrates.	Peckarsky (1979, 1980)	Retrieved using solid steel baffle over end of box to minimize loss of animals.
7. NATURAL ORGANIC SUBSTRATES		
a) *Leaf and needle packs*[d]		
Mesh bags (3.0-mm or 0.27-mm mesh size) with internal 1-cm mesh plastic support and filled with 20 g of oven-dried oak, sycamore or willow leaves.[a]	Mathews and Kowalczewski (1969)	No anchoring method described.
Leaf packs (10 g) of oven-dried leaves of a single species fastened together with plastic buttoneers and tied with nylon monofilament line to ~1-kg brick (i.e., mesh bags not	Petersen and Cummins (1974)	To remove, leaf pack was cut from brick and placed in a plastic bag.

Table I, continued

Description	Reference	Comment
used). Positioned with leaves on top, facing into current.[a]		
Leaves on needles air-dried and strung on monofilament line to produce leaf packs (5–15 g). Packs oven-dried and tied to upstream edge of bricks.[a]	Sedell et al. (1975)	
Bags of oven-dried needles and leaves fixed inside black plastic pipes (30 cm long, 51 mm ID). Pipes wired to plywood sheet held in place on stream bottom with rocks. Boxes also tried as containers for leaf packs.	Eidt and Meating (1978)	Pipes preferred over boxes. To retrieve, downstream end of pipe covered with a fine mesh net and the pipe containing the leaf pack removed.
Use of in-stream plastic cages to enclose leaf packs (Peterson and Cummins 1974) and either exclude or enclose selected organisms.	Merritt et al. (1979)	Mesh size of cages depends on size of organisms to be enclosed or selectively admitted and concentrated suspended particulates in transport.
b. *Miscellaneous*		
Mats of palm leaves usually enclosed by a frame or in a bag.	Azim and Ayyad (1948)[b] Stephenson (1947)[b] Marill (1958)[b] Chu and Vanderburg (1976)	Appears to have been used for snails only but has potential for other taxa.
Wooden sticks (50 cm long × 3 cm diameter) joined end to end to form a chain.[a]	McLachlan (1970)	Useful for studying colonization in flooded forest areas, etc.
White painted strips of bamboo. Length of cord suspended from bridge with mango leaves tied on at 50 cm intervals.[a,c]	Disney (1972)	
Barkless wooden logs (23 cm long × 6 cm diameter) set in strings up to 30 logs long.[a]	Nilsen and Larimore (1973)	

Table I, continued

Description	Reference	Comment
Cut stems of *Juncus effusus,* ends clamped by angle bracket placed over pegs set into stream bed.[a,c]	Wotton (1977)	
Lengths of fishing line, *Pandanus* leaves and wood of standard size.[a]	Hall and Edwards (1978)	No anchoring method described.
8. MISCELLANEOUS		
Pieces of tire (25 cm^2) cut from artificial reef formed from piles of automobile tires.[a]	Alfieri (1975)	Scuba used to collect sample.
Pieces of filter paper (20 × 20 cm) used as an attractant for snails.	Tanaka et al. (1975)	Not recommended except in very special situations, e.g., very shallow quiet water.
Description of a large number of anchored or floating devices.[a]	Fredeen and Spurr (1978)	Includes description of several anchoring methods.[c] Sampler not contained until they reached surface of water, if at all.
Suspended combination multiplate and basket samplers using a variety of materials as plates and as fill for baskets.[a]	Leglize (1978)	Diversity of fauna on the various materials used compared.

[a] No effective method described to prevent losses of animals on retrieval.
[b] Actual paper not read.
[c] Used mainly in work on simuliids.
[d] Since leaf pack and needle string sampling methods have developed apart from other artificial substrate sampling methods, only the initial and some key references are given.

Table II. Studies Comparing Different Types of Artificial Substrates and Comparing Arificial Substrates to Other Sampling Methods

	Containers – Rock-Filled	Containers – Other Fill	Multiplates	Panels, Tiles	Bricks, Blocks, etc.	Plastic Sheets, Polyethylene and Fabric Strips, Ropes, etc.	Implanted Substrates	Natural (organic) Substrates	Miscellaneous Artificial Substrates
Containers – rock-filled	Khalaf and Tachet (1980) Sheldon (1977) Townsend and Hildrew (1976) Moon (1940) Minshall and Minshall (1977) Zillich (1967) Wene and Wickliff (1940) Kreis et al. (1971) Bournaud et al. (1978) Wise & Molles (1979)								
Containers – other fill	Mason et al. (1973) Moon (1940) Zillich (1967) Bergersen and Galat (1975)	Benefield et al. (1974) Moon (1935, 1940) Armitage (1979)							

Table II, continued

	Containers—Rock-Filled	Containers—Other Fill	Multiplates	Panels, Tiles	Bricks, Blocks, etc.	Plastic Sheets, Polyethylene and Fabric Strips, Ropes, etc.	Implanted Substrates	Natural (organic) Substrates	Miscellaneous Artificial Substrates
	Voshell and Simmons (1977) Prins and Black (1971)								
Multiplates	Fullner (1971) Mason et al. (1973)	Mason et al. (1973)	Harrold (1978)						
Panels, tiles			Pardue (1973)	Markosova (1980) Konstantinov (1977)					
Bricks, blocks, etc.				Markosova (1980)	Markosova (1974) Hoar and Miller (1972)				
Plastic Sheets, polyethylene and fabric strips, ropes, etc.	Pearson and Jones (1975) Freeden and Spurr (1978)		Fredeen and Spurr (1978)	Gersabeck and Merritt (1979) Marsosova (1980) Boobar and Granett (1978)	Markosova (1980)	Markosova (1980) Fredeen and Spurr (1978) Elsen and Hebrard (1977) Elsen (1977) Pegel and Rühm (1976) Macan and Kitching (1972, 1976)			

Implanted substrates							Peckarsky (1980) Hynes (1974)		
Natural (organic) substrates						Boobar and Granett (1978)		Disney (1972) Nilsen and Larimore (1973) McLachlan (1970) Konstantinov (1977) Eidt and Meating (1978)	
Miscellaneous artificial substrates	Crowe (1974) Crossman and Cairns (1974) Freeden and Spurr (1978)	Crowe (1974)	Crowe (1974) Jarrett et al. (1975)			Williams and Obeng (1962) Fredeen and Spurr (1978) Tarshis (1968)			Nilsson and Sjöström (1977) Nakamura et al. (1978) Bourget and Lacroix (1971) Fredeen and Spurr (1978) Wolfe and Peterson (1958) Leglize (1978)
Hess	Minshall and Minshall (1977) Mason (1976) Rabeni and Gibbs (1978)					Boobar and Granett (1978)			

Table II, continued

	Containers–Rock-Filled	Containers–Other Fill	Multiplates	Panels, Tiles	Bricks, Blocks, etc.	Plastic Sheets, Polyethylene and Fabric Strips, Ropes, etc.	Implanted Substrates	Natural (organic) Substrates	Miscellaneous Artificial Substrates
Surber	Khalaf and Tachet (1980) Crossman and Cairns (1974) Pearson and Jones (1975) Townsend and Hildrew (1976) Friberg et al. (1977) Zillich (1967) Stanford (1972) Roux et al. (1976)	Jacobi (1971) Simmons and Winfield (1971)				Pearson and Jones (1975)	Hughes (1975) Radford and Hartland-Rowe (1971) Stanford and Reed (1974)		
Kick samples	Crossman and Cairns (1974) Hilsenhoff (1969) Minshall and Andrews (1973)	Benfield at al. (1974) Stauffer et al. (1976) Roby et al. (1978) Simmons and Winfield (1971)					Coleman and Hynes (1970) Egglishaw (1964)		

Grabs	Voshell and Simmons (1977) McConville (1975) Crowe (1971, 1972a) Anderson and Mason (1968)	McCall (1977) Voshell and Simmons (1977) Crowe (1971)	McConville (1975) McDaniel (1974) Tsui and Breedlove (1978) Cover and Harrel (1978)		McConville (1975)	McConville (1975) Barber et al. (1979)			
Other nonartificial substrate methods		Moon (1935, 1940) Dahlberg and Conyers (1973)		Lewis and Bennett (1974) Zahar (1951) Goddard et al. (1975) Friedrich and Herbst (1968)	Kovalak (1978)	Obeng (1967) Glime and Clemons (1972) Soszka (1975) Macan (1977) Gerrish and Bristow (1979) Pegel (1980)	Hughes (1975) Bishop (1973) Waters (1964)	Disney (1972) Chu and Vanderburg (1976)	Tanaka et al. (1975)

It is obvious from Table I that there is a very high diversity of kinds and modes of use of artificial substrate samplers, and of habitats in which they have been placed. Usage of a given sampler in a particular habitat will depend on the aims and objectives of the study to be undertaken, so no recommendations on usage can be given. A critical assessment of this and other aspects of the use of artificial substrates in the study of freshwater benthic macroinvertebrates is given in Chapter 6.

ACKNOWLEDGMENTS

D. G. Cobb and A. P. Weins helped produce the manuscript. D. Laroque and G. Decterow typed the drafts. K. E. Marshall did the literature search.

REFERENCES

Alfieri, D. J. "Organismal Development on an Artificial Substrate 1 July 1972–6 June 1974," *Eust. Coast. Mar. Sci.* 3: 465–472 (1975).

An der Lan, H. "Zur turbellarien–Fauna des hyporheischen Interstitials," *Arch. Hydrobiol., Suppl.* 33: 63–72 (1967).

Anderson, J. B., and W. T. Mason, Jr. "A Comparison of Benthic Macroinvertebrates Collected by Dredge and Basket Sampler," *J. Wat. Poll. Control Fed.* 40: 252–259 (1968).

Armitage, P. D. "A Folding Artificial Substratum Sampler for Use in Standing Water," *Hydrobiologia* 66: 245–248 (1979).

Arthur, J. W., and W. B. Horning, II. "The Use of Artificial Substrates in Pollution Surveys," *Am. Midl. Nat.* 82: 83–89 (1969).

Azim, M. A., and N. Ayyad. "A Preliminary Report on the Value of Palm-Leaf Traps in the Survey and Treatment of Streams Infested with Snails," *Trans. Roy. Soc. Trop. Med. Hyg.* 42: 231–246 (1948).

Barber, W. E., J. G. Greenwood and P. Crocos. "Artificial Seagrass–a New Technique for Sampling the Community," *Hydrobiologia* 65: 135–140 (1979).

Beak, T. W., T. C. Griffing and A. G. Appleby. "Use of Artificial Substrate Samplers to Assess Water Pollution," in *Biological Methods for the Assessment of Water Quality,* J. Cairns, Jr., and K. L. Dickson, Eds, *ASTM Tech Publ.* 528 (Philadelphia: American Society for Testing and Materials, 1973), pp. 227–241.

Benfield, E. F., A. C. Hendricks and J. Cairns, Jr. "Proficiencies of Two Artificial Substrates in Collecting Stream Macroinvertebrates," *Hydrobiologia* 45: 431–440 (1974).

Bergersen, E. P., and D. L. Galat. "Coniferous Tree Bark: a Lightweight Substitute for Limestone Rock in Barbeque Basket Macroinvertebrate Samplers," *Water Res.* 9: 729–731 (1975).

Besch, W., W. Hofmann and W. Ellenberger. "Das Makrobenthos auf Polyäthylensubstraten in Fliessgewässern. 1. Die Kinzig, ein Fluss der unteren Salmoniden und oberen Barbenzone," *Ann. Limnol.* 3: 331-368 (1967).

Bishop, J. E. "Observations on the Vertical Distribution of the Benthos in a Malaysian Stream," *Freshwater Biol.* 3: 147-156 (1973).

Boobar, L. R., and J. Granett. "Evaluation of Polyethylene Samplers for Black Fly Larvae (Diptera:Simuliidae), with Particular Reference to Maine Species," *Can. J. Zool.* 56: 2245-2248 (1978).

Bournaud, M., G. Chavanon and H. Tachet. "Structure et fonctionnement des écosystèmes du Haut-Rhône français. 5. Colonisation par les macroinvertébrés de substrats artificiels suspendus en pleine eau ou posés sur le fond," *Int. Ver. Theor. Angew. Limnol. Verh.* 20: 1485-1493 (1978).

Bourget, E., and G. Lacroix. "Two Simple Durable Epifaunal Collectors," *J. Fish. Res. Bd. Can.* 28: 1205-1207 (1971).

Britt, N. W. "New Methods of Collecting Bottom Fauna from Shoals or Rubble Bottom of Lakes and Streams," *Ecology* 36: 524-525 (1955).

Bull, C. J. "A Bottom Fauna Sampler for Use in Stony Streams," *Prog. Fish Cult.* 30: 119-120 (1968).

Chu, K. Y., and J. A. Vanderburg. "Techniques for Estimating Densities of *Bulinus truncatus rohlfsi* and its Horizontal Distribution in Volta Lake, Ghana," *Bull. World Health Org.* 54: 411-416 (1976).

Coleman, M. J., and H. B. N. Hynes. "The Vertical Distribution of the Invertebrate Fauna in the Bed of a Stream," *Limnol. Oceanogr.* 15: 31-40 (1970).

Cory, R. L. "Epifauna of the Patuxent River Estuary Maryland, for 1963 and 1964," *Chesapeake Sci.* 8: 71-89 (1967).

Cover, E. C., and R. C. Harrel. "Sequences of Colonization, Diversity, Biomass, and Productivity of Macroinvertebrates on Artificial Substrates in a Freshwater Canal," *Hydrobiologia* 59: 81-95 (1978).

Crandall, M. E. "Epibenthic Invertebrates of Croton Bay in the Hudson River," *New York Fish Game J.* 24: 178-186 (1977).

Crossman, J. S., and J. Cairns, Jr. "A Comparative Study Between Two Different Artificial Substrate Samplers and Regular Sampling Techniques," *Hydrobiologia* 44: 517-522 (1974).

Crowe, J.—A.M.E. "The Use of Two Types of Artificial Substrates to Sample the Macroinvertebrates of the Rat River, 1970." Manitoba Department of Mines and Natural Resources, Winnipeg, Manitoba, paper presented at 19th annual meeting of the Midwest Bethological Society, March 24-26, 1971, Purdue University, Purdue, IN, 1971.

Crowe, J.—A.M.E. "The Use of Rock-Filled Basket Samples to Survey the Aquatic Invertebrates of the Saskatchewan River," *Man. Dept. Mines Resources Environ. Managmt. Res. Branch MS Rep.* 72-73 (1972a).

Crowe, J.—A.M.E. "Saskatchewan River Survey, 1971," *Man. Dept. Mines Resources Environ. Managmt. Res. Branch MS Rep.* 72-77 (1972b).

Crowe, J.—A.M.E. "Tests of Artificial Samplers for Collecting Stream Macroinvertebrates in Manitoba," *Man. Entomol.* 8: 19-31 (1974).

Curtis, L. C. "A Method for Accurate Counting of Blackfly Larvae (Diptera:Simuliidae)," *Mosq. News.* 28: 238-239 (1968).

Dahlberg, M. D., and J. C. Conyers. "Winter Fauna in a Thermal Discharge with Observations on a Macrobenthos Sampler," in *Thermal Ecology.*, proceedings of a symposium held at Augusta, Georgia, May 3-5, 1973, U.S. Atomic Energy Commission Technical Information Center, Oak Ridge. J. W. Gibbons and R. R. Scharitz, eds. (1974), pp. 414-422.

Deutsch, W. G. "Macroinvertebrate Colonization of Acrylic Plates in a Large River," *Hydrobiologia* 75: 65-72 (1980).

Dickson, K. L., and J. Cairns, Jr. "The Relationship of Fresh-Water Macroinvertebrate Communities Collected by Floating Artificial Substrates to the MacArthur-Wilson Equilibrium Model," *Am. Midl. Nat.* 88: 68-75 (1972).

Dickson, K. L., J. Cairns, Jr. and J. C. Arnold "An Evaluation of the Use of a Basket-Type Artificial Substrate for Sampling Macroinvertebrate Organisms," *Trans. Am. Fish. Soc.* 100: 553-559 (1971).

Disney, R. H. L. "Observations on Sampling Pre-Imaginal Populations of Blackflies (Dipt., Simuliidae) in West Cameroon," *Bull. Entomol. Res.* 61: 485-503 (1972).

Doby, J. M., B. Rault and F. Beaucournu-Saguez. "Utilisation de rubans de plastique pour la récoleté des oeufs et des stades larvaires et nympheaux de simulies (Diptères Paranématocères) et pour l'étude biologique de ceux-ci." *Ann. Parasitol. Hum. Comp. 42: 651*-657 (1967).

Ebert, R. R. "A Study of the Communities of Aquatic Macroinvertebrates at the Merging of the Mississippi and Chippewa Rivers, Using a Biplate Substrate Sampler," M.Sc. Thesis, St. Mary's College, Winona, MN 66 (1965).

Egglishaw, H. J. "The Distributional Relationship Between the Bottom Fauna and Plant Detritus in Streams," *J. Animal Ecol.* 33: 463-476 (1964).

Eidt, D. C., and J. H. Meating. "Method for Incubating Leaf Tissue to Study Decomposition in Streams," *J. Fish. Res. Bd. Can.* 35: 247-248 (1978).

Elsen, P. "Méthodes d'échantillonnage des populations préimaginales de *Simulium damnosum* Theobald, 1903 (Diptera, Simuliidae) en Afrique de l"Ouest. I. Distribution verticale des larves et des nymphes: observations préliminaires," *Tropenmed. Parasitol.* 28: 91-96 (1977).

Elsen, P., and G. Hébrard. "Méthodes d'échantillonnage des populations préimaginales de *Simulium damnosum* Theobald, 1903 (Diptera, Simuliidae) en Afrique de l'Ouest. II. Observations sur le choix des couleurs, l'évolution du peuplement et la répartition horizontale au moyen de rubans en plastique," *Tropenmed. Parasitol.* 28: 471-477 (1977).

Flannagan, J. F., B. E. Townsend and B. G. E. de March. "Acute and Long-Term Effects of Methoxychlor Larviciding on the Aquatic Invertebrates of the Athabasca River, Alberta," in *Control of Blackflies in the Athabasca River,* W. W. Haufe and G. C. R. Croome, Eds. Technical Report, Alberta Department Environment, Edmonton, Alberta (1980), pp. 151-168.

Fredeen, F. J. H., and D. T. Spurr. "Collecting Semi-quantitative Samples of Black Fly Larvae (Diptera:Simuliidae) and Other Aquatic Insects from Large Rivers with the Aid of Artificial Substrates," *Quaest. Entomol.* 14: 411-431 (1978).

Friberg, F., L. M. Nilsson, C. Otto, P. Sjöström, B. W. Svensson, Bj. Svensson and S. Ulfstrand. "Diversity and Environments of Benthic Invertebrate Communities in South Swedish Streams," *Arch. Hydrobiol.* 81: 129-154 (1977).

Friedrich, G., and H. V. Herbst. "Der Litoralzylinder, ein Gerät zur quantitativen Probenentnahme im Uferbereich," *Gewäss. Abwäss.* 47: 79-87 (1968).

Fullner, R. W. "A Comparison of Macroinvertebrates Collected by Basket and Modified Multiple-Plate Samplers," *J. Water Poll. Control Fed.* 43: 494-499 (1971).

Gerrish, N., and J. M. Bristow. "Macroinvertebrate Associations with Aquatic Macrophytes and Artificial Substrates," *J. Great Lakes Res.* 5: 69-72 (1979).

Gersabeck, E. F., Jr., and R. W. Merritt. "The Effect of Physical Factors on the Colonization and Relocation Behavior of Immature Black Flies (Diptera: Simuliidae)," *Environ. Entomol.* 8: 34-39 (1979).

Gittins, B. T. "A Littoral Artificial Substrate," *Br. Phycol. J.* 11: 383-386 (1976).

Glime, J. M., and R. M. Clemons. "Species Diversity of Stream Insects on *Fontinalis* spp. Compared to Diversity on Artificial Substrates," *Ecology* 53: 458-464 (1972).

Goddard, C. I., M. H. Goodwin and L. A. Greig. "The Use of Artificial Substrates in Sampling Estuarine Benthos," *Trans. Am. Fish. Soc.* 104: 50-52 (1975).

Hall, R. O., and A. J. Edwards. "Observations on the Settling of *Simulium damnosum* Larvae on Artificial Substrates in the Ivory Coast," *Hydrobiologia* 57: 81-84 (1978).

Harrold, J. F., Jr. "Relation of Sample Variations to Plate Orientation in the Hester-Dendy Plate Sampler," *Prog. Fish Cult.* 40: 24-25 (1978).

Herrmann, J. "Population Dynamics of *Dendrocoelum lacteum* (O. F. Müller) (Turbellaria, Tricladida) in a South Swedish Lake," *Arch. Hydrobiol.* 85: 482-510 (1979).

Hester, F. E., and J. S. Dendy. "A Multiple-Plate Sampler for Aquatic Macroinvertebrates," *Trans. Am. Fish. Soc.* 91: 420-421 (1962).

Hilsenhoff, W. L. "An Artificial Substrate Device for Sampling Benthic Stream Invertebrates," *Limnol. Oceanog.* 14: 465-471 (1969).

Hoar, M. A., and M. C. Miller. "Midsummer Studies of a New Year-Round Artificial Substrate for Macroinvertebrate Populations in the Upper Mississippi River," M.Sc. Thesis, St. Mary's College, Winona, MN (1972).

Hocutt, C. H., K. L. Dickson and M. T. Masnik. "Methodology Developed for Sampling Macroinvertebrates by Artificial Substrates in the New River, Virginia," *Rev. Biol.* (Lisb.) 10: 63-75 (1974-1976).

Hughes, B. D. "A Comparison for Four Samplers for Benthic Macroinvertebrates Inhabiting Coarse River Deposits," *Water Res.* 9: 61-69 (1975).

Hynes, H. B. N. "Further Studies on the Distribution of Stream Animals Within the Substratum." *Limnol. Oceanog.* 19: 92-99 (1974).

Jacobi, G. Z. "A Quantitative Artificial Substrate Sampler for Benthic Macroinvertebrates," *Trans. Am. Fish. Soc.* 100: 136-138 (1971).

Jarrett, F. L., K. B. Grogan, D. L. Martin and J. W. McIntosh, Jr. "Use of Artificial Substrate for Sampling Macroinvertebrate Organisms," *Va. J. Sci.* 26: 56 (1975).

Kathman, D. "Artificial Substrate Sampler for Benthic Invertebrates in Ponds, Small Lakes and Reservoirs," *Prog. Fish Cult.* 40: 114-115 (1978).

Khalaf, G., and H. Tachet. "La dynamique de colonisation des substrats artificiels par les macroinvertébré d'un cours d'eau," *Ann. Limnol.* 13: 169-190 (1977).

Khalaf, G., and H. Tachet. "Colonization of artificial substrata by macroinvertebrates in a stream and variations according to stone size," *Freshwater Biol.* 10: 475-482 (1980).

Konstantinov, A. S. "Ability of Chironomid Larvae to Colonize Substrates Variously Positioned in a Lake," *Hydrobiol. J.* 13: 11-14 (1977).

Kovalak, W. P. "Seasonal and Diel Changes in the Positioning of *Glossosoma nigrior* Banks (Trichoptera: Glossosomatidae) on Artificial Substrates," *Can. J. Zool.* 54: 1585-1594. (1976).

Kovalak, W. P. "Diel Changes in Stream Benthos Density on Stones and Artificial Substrates," *Hydrobiologia* 58: 7-16 (1978).

Kovalak, W. P. "Day-Night Changes in Stream Benthos Density in Relation to Current Velocity," *Arch. Hydrobiol.* 87: 1-18 (1979).

Kreis, R. D., and R. L. Smith. "A Method of Suspending Multiple "Basket Samplers" in Reservoirs," *Prog. Fish Cult.* 32: 182-183 (1970).

Kreis, R. D., R. L. Smith and J. E. Moyer. "The Use of Limestone-Filled Basket Samplers for Collecting Reservoir Macroinvertebrates," *Water Res.* 5: 1099-1106 (1971).

Lapchin, L. "Utilisation de substrats artificiels pour l'étude des populations d'invertébrés benthiques. Résultats préliminaires dans un ruisseau à Salmonidés de Bretagne," *Ann. Hydrobiol.* 8: 33-44 (1977).

Laville, H. "Utilisation de substrats artificiels pour l'étude de la faune macrobenthique de la zone littorale rocheuse des lacs de montagne," *Ann. Limnol.* 10: 163-172. (1974).

Legier, P., and J. Talin. "Recolonisation d'un ruisseau temporaire et evolution du degré de stabilité de la zoocénose," *Ecol. Mediterr.* 1: 149-164 (1975).

Leglize, L. "Approche méthodologique de la faune benthique de la Moselle: description d'un substrat artificiel composite et résultats préliminaires," *Cahiers Lab. Hydrobiol. Montereau* 7: 61-65 (1978).

Lewis, D. J., and G. F. Bennett. "An Artificial Substrate for the Quantitative Comparison of the Densities of Larval Simuliid (Diptera) Populations," *Can. J. Zool.* 52: 773-775 (1974).

Macan, T. T. "The Fauna in the Vegetation of a Moorland Fishpond as Revealed by Different Methods of Collecting," *Hydrobiologia* 55: 3-15 (1977).

Macan, T. T., and A. Kitching. "Some Experiments with Artificial Substrata," *Int. Ver. Theor. Angew Limnol. Verh.* 18: 213-220 (1972).

Macan, T. T., and A. Kitching. "The Colonization of Squares of Plastic Suspended in Mid-Water," *Freshwater Biol.* 6: 33-40 (1976).

Marill, F. G. "Sur l'appréciation comparative de la richesse des gîtes en mollusque fluviatiles notamment en *Bulinus truncatus* Audouin," *Bull. World Health* 18: 1057-1064 (1958).

Markosova, R. "Seasonal Dynamics of the Periphytic Macrofauna in Carp Ponds in South-west Bohemia," *Vestn. Cesk. Spol. Zool.* 38: 251-270 (1974).

Markosova, R. "Develoment of the Periphytic Community on Artificial Substrates in Fish Ponds,' *Int. Rev. Ges. Hydrobiol.* 64: 811-825 (1979).

Markosova, R. "Effect of Submerged Substrates Quality on the Periphytic Macrofauna in Ponds," *Vestn. Cesk. Spol. Zool.* 44: 46-52 (1980).

Mason, J. C. "Evaluating a Substrate Tray for Sampling the Invertebrate Fauna of Small Streams, with Comment on General Sampling Problems," *Arch. Hydrobiol.* 78: 51-70 (1976).

Mason, W. T., Jr., J. B. Anderson, R. D. Kreis and W. C. Johnson. "Artificial Substrate Sampling, Macroinvertebrates in a Polluted Reach of the Klamath River, Oregon," *J. Water Poll. Control Fed.* 42: R315-R328. (1970).

Mason, W. T., J. B. Anderson and G. E. Morrison. "A Limestone-Filled, Artificial Substrate Sampler-Float Unit for Collecting Microinvertebrates in Large Streams,' *Prog. Fish Cult.* 29: 74 (1967).

Mason, W. T., Jr., C. I. Weber, P. A. Lewis and E. C. Julian. "Factors Affecting the Performance of Basket and Multiplate Macroinvertebrate Samplers," *Freshwater Biol.* 3: 409-436. (1973).

Mathers, C. K., and T. R. Martin. "A New Multiple-Plate Sampler for Collecting Macroinvertebrates of the Stream Biocies," *Trans. Ill. State Acad. Sci.* 62: 331-333 (1969).

Mathews, C. P., and A. Kowalczewski. "The Disappearance of Leaf Litter and its Contribution to Production in the River Thames," *J. Ecol.* 57: 543-552 (1969).

McCall, P. L. "Community Patterns and Adaptive Strategies of the Infaunal Benthos of Long Island Sound," *J. Mar. Res.* 35: 221-266 (1977).

McConville, D. R. "Comparison of Artificial Substrates in Bottom Fauna Studies on a Large River," *J. Minn. Acad. Sci.* 41: 21-24 (1975).

McDaniel, M. D. "Design and Preliminary Evaluation of an Improved Artificial Substrate Sampler for Aquatic Macroinvertebrates," *Prog. Fish Cult.* 36: 23-25 (1974).

McLachlan, A. J. "Submerged Trees as a Substrate for Benthic Fauna in the Recently Created Lake Kariba (Central Africa)," *J. Appl. Ecol.* 7: 253-266 (1970).

Meier, P. G., D. L. Penrose and L. Polak. "The Rate of Colonization by Macroinvertebrates on Artificial Substrate Samplers," *Freshwater Biol.* 9: 381-392 (1979).

Merritt, R. W., K. W. Cummins and J. R. Barnes. "Demonstration of Stream Watershed Community Processes with Some Simple Bioassay Techniques," in *Innovative Teaching in Aquatic Entomology,* V. H. Resh and D. M. Rosenberg, Eds. *Can. Spec. Publ. Fish. Aquat. Sci.* 43: 101-113 (1979).

Minshall, G. W., and D. A. Andrews. "An Ecological Investigation of the Portneuf River, Idaho: a Semiarid-Land Stream Subjected to Pollution," *Freshwater Biol.* 3: 1-30 (1973).

Minshall, G. W., and J. N. Minshall. "Microdistribution of Benthic Invertebrates in a Rocky Mountain (U.S.A.) Stream," *Hydrobiologia* 55: 231-249 (1977).

Moon, H. P. "Methods and Apparatus Suitable for an Investigation of the Littoral Region of Oligotrophic Lakes," *Int. Rev. Ges. Hydrobiol. Hydrogr.* 32: 319-333 (1935).

Moon, H. P. "An Investigation of the Movements of Freshwater Invertebrate faunas," *J. Animal Ecol.* 9: 76-83 (1940).

Moore, R. L., and S. F. Cook, Jr. "Pan Traps for Migration Studies of Chaoborid Midge Larvae (Diptera: Culicidae)," *Ann. Entomol. Soc. Am.* 60: 478 (1967).

Mundie, J. H. "A Bottom Sampler for Inclined Rock Surfaces in Lakes," *J. Animal Ecol.* 25: 429–432 (1956).

Nakamura, Y., K. Saito and M. Takahashi. "Studies on the Ecology of Black Flies (Simuliidae:Diptera). (1) Comparison of Some Quantitative Sampling Methods for Larvae," *Japan J. Sanit. Zool.* 29: 209–212 (1978) (Japanese with English summary)

Nilsen, H. C., and R. W. Larimore. "Establishment of Invertebrate Communities on Log Substrates in the Kaskaskia River, Illinois," *Ecology* 54: 366–374 (1973).

Nilsson, L. M., and P. Sjöström. "Colonization of Implanted Substrates by Differently Sized *Gammarus pulex* (Amphipoda)," *Oikos* 28: 43–48 (1977).

Obeng, L. E. "Life-History and Population Studies on the Simuliidae of North Wales," *Ann. Trop. Med. Parasitol.* 61: 472–487 (1967).

Pardue, G. P. "Production Response of the Bluegill Sunfish, *Lepomis macrochirus* Rafinesque, to Added Attachment Surface for Fish-Food Organisms," *Trans. Am. Fish. Soc.* 102: 622–626 (1973).

Pardue, G. P., and L. A. Nielsen. "Invertebrate Biomass and Fish Production in Ponds with Added Attachment Surface," in *Response of Fish to Habitat Structure in Standing Water,* D. L. Johnson and R. A. Stein, Eds., Am. Fish Soc. North Cent. Div. Spec. Publ. 6 (1979) pp. 34–37.

Parsons, D. A., and J. W. Tatum. "A New Shallow Water Multiple-Plate Sampler," *Prog. Fish Cult.* 36: 179–180 (1974).

Pearson, R. G., and N. V. Jones. "The Colonization of Artificial Substrata by Stream Macroinvertebrates," in *7th Int. Conf. on Water Poll. Res., Paris,* September 9-13, 1974, Vol. 1. S. H. Jenkins, Ed. *Prog. Water Technol.* 497-504 (1975).

Peckarsky, B. L. "Biological Interactions as Determinants of Distributions of Benthic Invertebrates Within the Substrate of Stony Streams," *Limnol. Oceanog.* 24: 59–68. (1979).

Peckarsky, B. L. "Influence of Detritus upon Colonization of Stream Invertebrates," *Can. J. Fish. Aquat. Sci.* 37: 957-963 (1980).

Pegel, M. "Zur Methodik der Driftmessung in der Fliessgewässerökologie unter besonderer Berücksichtigung der Simuliidae (Diptera)," *Z. Angew. Entomol.* 89: 198–214 (1980).

Pegel, M., and W. Rühm. "Versuche zur Besiedlung künstlicher Substrate durch präimaginale Stadien von Simuliiden unter besonderer Berücksichtigung von *Boophthora erythrocephala* de Geer (Simuliidae, Dipt.)," *Z. Angew. Entomol.* 82: 65–71 (1976).

Petersen, R. C., and K. W. Cummins. "Leaf Processing in a Woodland Stream," *Freshwater Biol.* 4: 343-368 (1974).

Prins, R., and W. Black. "Synthetic Webbing as an Effective Macrobenthos Sampling Substrate in Reservoirs," in *Reservoir Fisheries and Limnology,* G. E. Hall, Ed. *Am. Fish. Soc. Spec. Publ.* 8: 203-208 (1971).

Protasov, A. A. "Dynamics of Zooperiphyton Species of Experimental Substrates when Affected by the Heated Waters of Thermal Power Plants," *Hydrobiol. J.* 15: 41–43 (1979).

Rabeni, C. F., and K. E. Gibbs. "Comparison of Two Methods Used by Divers for Sampling Benthic Invertebrates in Deep Rivers," *J. Fish. Res. Bd. Can.* 35: 332–336 (1978).

Radford, D. S., and R. Hartland-Rowe. "Subsurface and Surface Sampling of Benthic Invertebrates in Two Streams," *Limnol. Oceanog.* 16: 114–120 (1971).

Roby, K. B., J. D. Newbold and D. C. Erman. "Effectiveness of an Artificial Substrate for Sampling Macroinvertebrates in Small Streams," *Freshwater Biol.* 8: 1–8 (1978).

Roux, A. L., H. Tachet and M. Neyron. "Structure et fonctionnement des écosystèmes du Haut-Rhone Francais. III. Une technique simple et peu onéreuse pour l'étude des macroinvertébrés benthiques des grands fleuves," *Bull. Ecol.* 7: 493–496 (1976).

Schwoerbel, J. "Das hyporheische Interstitial als Grenzbiotop zwischen oberirdischem und subterranen Ökosystem und seine Bedeutung für die Primär-Evolution von Kleinsthöhlenbewohnern," *Arch. Hydrobiol. Suppl.* 33: 1–62 (1967).

Scott, D. C. "Biological Balance in Streams," *Sewage Ind. Wastes* 30: 1169–1173 (1958).

Sedell, J. R., F. J. Triska and N. S. Triska. "The Processing of Conifer and Hardwood Leaves in Two Coniferous Forest Streams 1. Weight Loss and Associated Invertebrates," *Int. Ver. Theor. Angew. Limnol. Verh.* 19: 1617–1627 (1975).

Shaw, D. W., and G. W. Minshall. "Colonization of an Introduced Substrate by Stream Macroinvertebrates," *Oikos* 34: 259–271 (1980).

Sheldon, A. L. "Colonization Curves: Application to Stream Insects on Seminatural Substrates," *Oikos* 28: 256–261 (1977).

Shelford, V. E., and S. Eddy. "Methods for the Study of Stream Communities," *Ecology* 10: 382–391 (1929).

Simmons, G. M., Jr., and A. Winfield. "A Feasibility Study Using Conservation Webbing as an Artificial Substrate in Macrobenthic Studies," *VA. J. Sci.* 22: 52–59 (1971).

Soszka, G. J. "Ecological Relations Between Invertebrates and Submerged Macrophytes in the Lake Littoral," *Ekol. Pol.* 23: 393–415. (1975).

Stanford, J. "Comparison of Two Methods of Quantitatively Sampling Benthic Stream Macroinvertebrates," *J. Colo.-Wyo. Acad. Sci.* 7: 87 (1972).

Stanford, J. A., and E. B. Reed. "A Basket Sampling Techniques for Quantifying Riverine Macrobenthos," *Water Resources Bull.* 10: 470–477 (1974).

Stauffer, J. R., H. A. Beiles, J. W. Cox, K. L. Dickson and D. E. Simonet. "Colonization of Macrobenthic Communities on Artificial Substrates," *Rev. Biol.* (Lisb.) 10: 49–61 (1974–1976).

Stephenson, R. W. "Bilharziasis in the Gezira Irrigated Area of the Sudan," *Trans. R. Soc. Trop. Med. Hyg.* 40: 479–494 (1947).

Tanaka, H., M. J. Santos, H. Matsuda, K. Yasuraoka and A. T. Santos, Jr. "A Quantitative Sampling Method for *Oncomelania quadrasi* by Fiter Paper," *Japan J. Exp. Med.* 45: 255–262 (1975).

Tarshis, I. B. "Use of Fabrics in Streams to Collect Black Fly Larvae," *Ann. Entomol. Soc. Am.* 61: 960-961 (1968).

Townsend, C. R., and A. G. Hildrew. "Field Experiments on the Drifting, Colonization and Continuous Redistribution of Stream Benthos," *J. Animal Ecol.* 45: 759-772 (1976).

Tsui, P. T. P., and B. W. Breedlove. "Use of the Multiple-Plate Sampler in Biological Monitoring of the Aquatic Environment," *Fla. Sci.* 41: 110-116 (1978).

Ulfstrand, S., L. M. Nilsson and A. Stergar. "Composition and Diversity of Benthic Species Collectives Colonizing Implanted Substrates in a South Swedish Stream," *Entomol. Scand.* 5: 115-122 (1974).

Voshell, J. R., Jr., and G. M. Simmons, Jr. "An Evaluation of Artificial Substrates for Sampling Macrobenthos in Reservoirs," *Hydrobiologia* 53: 257-269 (1977).

Waters, T. F. "Recolonization of Denuded Stream Bottom Areas by Drift," *Trans. Am. Fish. Soc.* 93: 311-315 (1964).

Weber, C. I., ed. "Biological Field and Laboratory Methods for Measuring the Quality of Surface Waters and Effluents," *U.S. Environmental Protection Agency, Environmental Monitoring Service,* EPA-670/4-73-001 (1973).

Wefring, D. R., and J. C. Teed. "Device for Collecting Replicate Artificial Substrate Samples of Benthic Invertebrates in Large Rivers," *Prog. Fish Cult.* 42: 26-28. (1980).

Wene, G., and E. L. Wickliff. "Modification of a Stream Bottom and Its Effects on the Insect Fauna," *Can. Entomol.* 72: 131-135 (1940).

Williams, T. R., and L. Obeng. "A Comparison of Two Methods of Estimating Changes in *Simulium* Larval Populations with a Description of a New Method," *Ann. Trop. Med. Parasitol.* 56: 359-361 (1962).

Wise, D. H., and M. C. Molles, Jr. "Colonization of Artificial Substrates by Stream Insects: Influence of Substrate Size and Diversity," *Hydrobiologia* 65: 69-74 (1979).

Witham, R., R. M. Ingle and E. A. Joyce, Jr. "Physiological and Ecological Studies of *Panulirus argus* from the St. Lucie Estuary," *Fla. Bd. Conserv. Mar. Lab. Tech. Ser.* 53: 1-31 (1968).

Wolfe, L. S., and D. G. Peterson. "A New Method to Estimate Levels of Infestations of Black-Fly Larvae (Diptera:Simuliidae)," *Can. J. Zool.* 36: 863-867 (1958).

Wotton, R. S. "Sampling Moorland Stream Blackfly Larvae (Diptera:Simuliidae)," *Arch. Hydrobiol.* 79: 404-412 (1977).

Zahar, A. R. "The Ecology and Distribution of Blackflies (Simuliidae) in South-east Scotland," *J. Animal Ecol.* 20: 33-62 (1951).

Zillich, J. A. "Responses of Lotic Insects to Artificial Substrate Samplers," M.Sc. Thesis, University of Wisconsin (1967).

CHAPTER 8

PRIORITIES FOR MARINE DEVELOPMENTS IN ARTIFICIAL SUBSTRATE BIOASSAYS: SYMPOSIUM COMMENT

Derek V. Ellis

Biology Department
University of Victoria
Victoria, British Columbia

INTRODUCTION

One of the objectives of this symposium was to consider how developments in artificial substrate protocols can lead to standardized methods in environmental impact assessment. Some use of such substrate procedures has been made over the past decade in our marine area of the Pacific coast of Canada, primarily at mine and pulp mill discharges. Typically, the results of such assessments remain in limited distribution reports, and the data on which these comments are based have not yet reached the open literature. In general, the data lead to the conclusion that protocols for artificial substrate bioassays in the sea are undeveloped when compared to freshwater use, and presently there is virtually no regulatory agency interested in the concept that standardized artificial substrate tests could provide very useful multispecies bioassays. The following paragraphs provide documentation to support these contentions.

First of all, of course, there is a long history extending back at least to the 1930s [1] of the use of artificial substrate assays in the sea through anti-fouling experiments. This traditional use is designed to get ships moving faster while at sea (frictional resistance may increase ½% per day—[2] and to spend less time out of action in dry dock. However, there has been a very noticeable trend in recent years to use the methods in more academic ways and, like any pure research, may eventually provide the spinoff needed for beneficial applications. The more academic interests now in vogue are to use artificial substrates for experimental analyses of ecosystem processes, whether successional, resource partitioning or other theoretical constructs. The first two papers presented at this symposium, those by Schoener (Chapter 1) and Harris (Chapter 5), are typical of this use. It is a very powerful approach with the potential to control many of the multiple parameters inevitably affecting the natural environment in shallow seas and confusing their analysis.

However, such use is a long way from the development of standard bioassay methods, which must overcome a number of standardization problems that have nothing to do with the use of the procedures in analyzing site-specific functional aspects of ecosystems. There are at least three sets of problems that need to be overcome if the assays are to be developed for widespread use.

RESPONSE PARAMETERS

Conventionally, response parameters used in artificial substrate experiments have been single-number measures such as percent cover by sessile organisms, numbers of species or more complex diversity indices. Table I, from an artificial substrate experiement arrayed down-fjord from a pulp mill discharge (conducted by the mill and analyzed by S. McKinnell), shows how misleading such single-measure numbers can be. Shannon-Weaver (H") and Evenness (J) indices (Figure 1) show levels unrelated to distance from the discharge, other than lowest diversity at the most distant station (# 6), which is due to the predominance of the colonial diatom *Amphipleura*. Table I shows that there are, nevertheless, very substantial species associational differences among all sites, which can be documented in various ways. We used the Czekanowski and Bray-Curtis similarity and dissimilarity measures. Almost all similarity measures were below 0.5 and almost all dissimilarity measures above 0.6. By the latter measure, stations 1 and 6 (the nearest and farthest from the discharge, respectively) were least dissimilar. We concluded that the

Table I. Specimen-Site Matrix Showing Proportions of Total Diatoms Represented by Each Species

Specimens	Stations					
	1	2	3	4	5	6
Opephora sp. A	0.0526		0.0310	0.0101		
Amphipleura sp. A	0.3083					0.8000
Navicula sp. A	0.0075					
Nitzschia sp. A	0.0075					
Nitzschia sp. B	0.0075	0.0441				
Nitzschia sp. D	0.0150					
Nitzschia sp. E	0.0227		0.0156	0.0606		
Nitzschia sp. F	0.0150					
Amphipleura sp. B	0.3083	0.2132	0.0233		0.0633	0.0756
Fragilaria pinnata	0.0451					
Acnanthes sp. A	0.0075					
Acnanthes sp. B	0.0075					
Acnanthes sp. C	0.0075					
Acnanthes sp. D	0.0075		0.0156			
Opephora sp. B	0.0075					
Genus I	0.0451					
Amphora sp. B	0.0376	0.0221	0.0310	0.2828	0.0063	0.0098
Amphora sp. A	0.0301	0.0074				
Navicula sp. B	0.0075					
Navicula sp. L	0.0150					
Cocconeis sp. A	0.0075					
Melosira sp. A	0.0226	0.0074	0.0620	0.0303		
Hantzschia sp. A	0.0075					
Navicula sp. M	0.0075					
Fragilaria sp. A		0.0074		0.0707	0.0253	
Fragilaria sp. B		0.0221		0.0505	0.0316	
Fragilaria sp. C		0.0735		0.0404	0.0127	
Fragilaria sp. E		0.0147	0.0156			
Fragilaria sp. F		0.0147		0.0303	0.1076	
Navicula sp. O		0.0294				
Navicula sp. D		0.0074	0.0156			
Navicula sp. N		0.0074				
Genus II		0.0147				0.0707
Thallasiothrix sp.		0.0588	0.0388	0.0404	0.0886	
Nitzschia sp. G		0.0074				
Nitzschia sp. C		0.0147	0.0233			
Acnanthes sp. E		0.0294				
Cocconeis sp. A		0.0074	0.0078			
Genus III		0.0221				
Cocconeis sp. B		0.0074		0.0101		
Cocconeis sp. C		0.0074		0.0101		
Amphipleura sp. C		0.0368				
Genus IV		0.0147				

Table I, continued

Specimens	Stations					
	1	2	3	4	5	6
Biddulphia sp. A		0.1385				0.0024
Centric A			0.0271			
Navicula sp. C			0.0310			
Navicula sp. E			0.0078			
Navicula sp. F			0.0078			
Centric B			0.0078			
Centric C			0.0233			
Centric D			0.1667			
Grammatophora sp.			0.0310			
Cocconeis sp. F			0.0156			
Cocconeis sp. D			0.0156			
Nitzschia sp. H			0.0233	0.0202	0.0063	0.0049
Nitzschia sp. I			0.0078			
Nitzschia sp. J			0.0078	0.0202	0.0063	0.0073
Nitzschia sp. K			0.0078			
Nitzschia sp. L			0.0078	0.0909	0.0190	0.0024
Genus V			0.0775	0.0303	0.0127	
Licormorpha sp. A			0.0078			
Tabellariaceae A			0.0078			
Amphora sp. C			0.0233			
Small Septate Naviculoid			0.0543	0.0303	0.0316	
Acnanthes sp. D			0.0156			
Amphora sp. D			0.0078			
Navicula sp. G				0.0202		
Acnanthes brevipes				0.0101		
Amphora sp. E				0.0202		
Hantzschia sp. C				0.0505		
Licomorpha sp. B				0.0101	0.0127	
Navicula sp. H				0.0101		
Rhoicosphenia curvate				0.0202		0.0049
Nitaschia sp. M				0.0101		
Hantzschia sp. D				0.0101		
Large Septate Naviculoid					0.0949	
Navicula sp. I					0.4304	
Navicula sp. J					0.0063	
Navicula sp. K					0.0127	
Cymbella sp. A					0.0253	
Thallasiosira sp. A					0.0063	
Cocconeis sp. G						0.0024
Navicula sp. P						0.0049
Amphora sp. F						0.0049
Hantzschia sp. F						0.0024

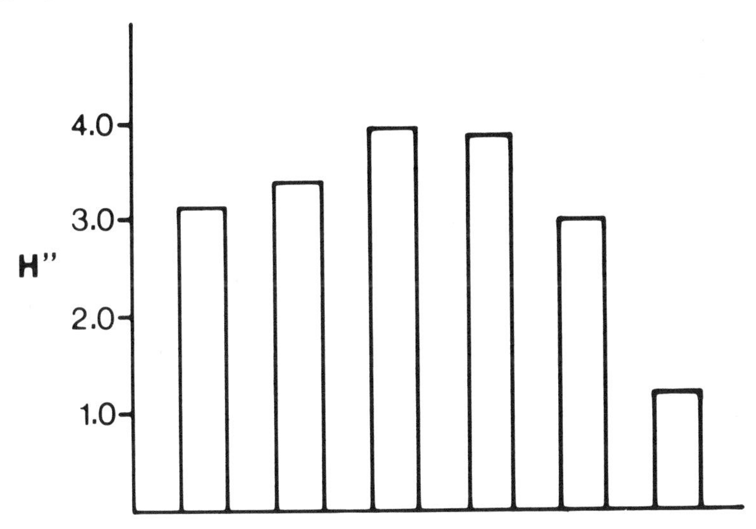

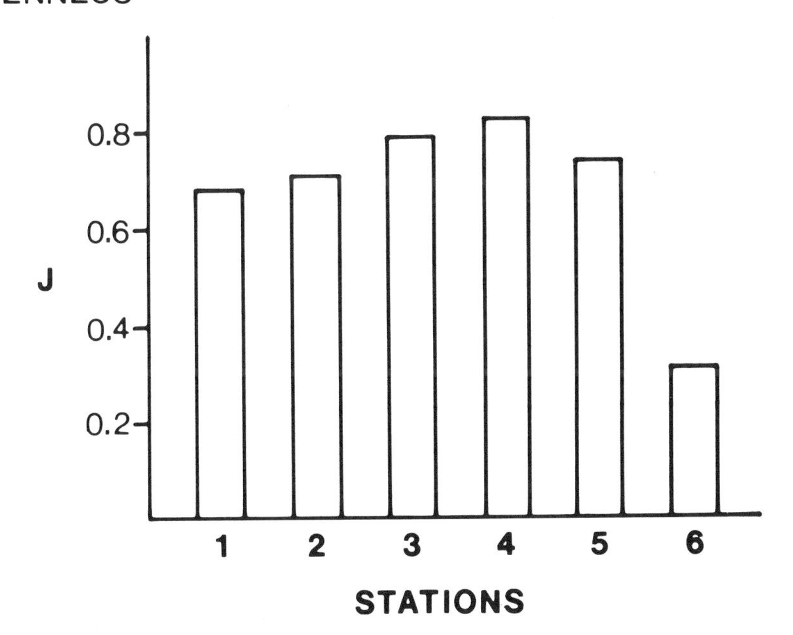

Figure 1. Species diversity and evenness.

variation was not due to the mill discharge and recommended additional substrates closer in.

In recent years an enormous array of multivariate statistical procedures for analyzing species associational data have been developed [3,4]. Artificial substrate assays must draw on these analytical methods when the responses are sufficiently subtle that they must be considered species associational changes rather than diversity losses. The problem is really to standardize decision-making on which of the two scales of response parameter (diversity or species associational) to use. I can see no alternative at this time in marine assays to site-specific pilot experiments, possibly for a full year or longer in duration, to indicate the diversity of the substrate community, the nature of any community equilibrium that may be achieved, the distance over which impacts grade into natural variation, and from these results conclude which scale of statistical analysis is needed, diversity or species association.

The only nonnumerical alternative to this complex statistical approach that I can see is broad-scale typology, such as documenting the presence-absence of such tax as "filamentous green algae," "barnacles," etc., or even amounts of "volatile organics," well documented by color photographic prints. Color prints can be assembled into very informative time-series matrices of stations by time for wall displays or even conventional reports on 8½ x 11 page format. This is one advantage of marine artificial substrate experiments as opposed to freshwater in that from photography many organisms are large and identifiable to their phyla, class, order and family, if not genus. I know of one such time-series still accumulating since 1971 at a mine tailing discharge site and available for inspection as a record of impact as needed.

TAXON IDENTIFICATIONS

Diversity index calculations have merit for cost-conscious environmental impact assessment (EIA) administrators in that they eliminate the expensive time requirement for biological species identifications. The more inherently diverse the biological community, the more time-consuming it will be to make accurate and precise identifications of biological species. The sea is populated by a range of organisms that spans far more phyla than those found in freshwater. Unfortunately, there is a problem in training identifiers not only in the technical proficiency of manipulating specimens from various phyla of solitary and colonial organisms and in acquiring the morphological jargon of each, but also in understanding the significance of differences in the concept of the species in the differently breeding phyla. Any laboratory engaged in routine biological species identifications is working in the area of traditional museum taxonomy and must have at least one professional biologist who is aware of, and sympathetic to, the classificatory and taxonomic problems that will arise when previously undescribed and unnamed species appear.

Table II documents the results of two interlaboratory calibrations of species identifications that have been made recently in our area. The procedure for specimen distribution was modelled on ring test procedures run by chemical analytical laboratories [5]. Part A spans seveal phyla, while Part B deals with a reasonably common group in EIA, the gammarid amphipoda. Notice the level of discrepancy between laboratories. In the multiphyla assembly (Part A), no one laboratory achieved 100%. In the assembly of amphipods, four of the participating laboratories achieved 100% agreement, with a progressive decline among the others to agreement in only one out of six. The fully agreeing laboratories included the two in the exercise that had an extensive professional publications record with the taxon and, thus, established the accuracy baseline against which the others were compared. This emphasizes that diversity and species associational comparisons between substrates are only as reliable as the raw identification data on which the statistical com-

Table II. Species Identification Intercalibrations

Part A. Several Phyla Assembly (arranged by C. Levings)

	Laboratories						
Taxa	**A**	**B**	**C**	**D**	**G**	**H**	**Major Groups**
Geodia mesotriaena	–	*	–	*	–	–	Sponge
Gammarus setosus	+	½ +	½ +	–	+	½ +	Amphipod
Orchestia traskiana	–	+	+	+	+	+	Amphipod
Neomysis awatschensis (= *N. mercedis*)	+	*	+	+	+	+	Mysid
Manayunkia aestruarina	½ +	*	½ +	+	+	+	Polychaete
Acila castrensis	+	*	+		+	+	Bivalve
Corophium insidiosum	½ +	+	+	+	+	+	Amphipod
Amphicteis	–	*	+	–	+	+	Polychaete
Barantolla americana	–	*	+	–	+	+	Polychaete

Legend
+ correctly identified to species
½ + correctly identified to genus
– incorrectly identified to genus and species
* no attempt made at identification

Part B. Gammarid Amphipod Assembly arranged by Cross and Ellis [6]

Taxa	Laboratories								
	B	**J**	**L**	**H**	**F**	**C**	**I**	**E**	**D**
Anisogammarus confervicolus	X	X	X	X	X	X	X	X	X
Orchestia traskiana	X	X	X	X	X	X			
Orchestia georgiana								X	
Orchestia sp.							X		
No ID									X
Allorchestes angusta	X	X	X	X	X	X			
Hyalella azteca								X	
Hyalella sp.							X		
Pontogeneia intermedia									X
Hyale grandicornis californica	X	X	X	X	X				
Hyale plumulosa								X	
Hyale sp.							X		
Orchestia georgiana						X			
No ID									X
Oligochinus lighti	X	X	X	X	X				
Paramoera carlottensis									X
Pontogeneia rostrata						X			
Pontogeneia sp.								X	
Calliopiella sp.							X		
Ampithoe simulans	X	X	X	X					
Ampithoe valida					X	X			
Amphithoe sp.							X	X	
No ID									X

parisons are based (whether sophisticated and computerized or not). Until species identifiers can obtain precise identifications, (i.e., specimens belonging to the same species are always assigned to the same species) and accurate identifications, (i.e., specimens are always assigned to the correct species if previously known to science and, if not previously known, recognized as new, provisionally designated, and set aside for description and definitive naming), then species associational and diversity comparisons between substrates are liable to be misleading. In this kind of work new species will be found and must be accommodated without reducing the accuracy of the statistics.

The approach of selecting from the substrate community only those taxa in which the investigator has identification capability, e.g., protozoa, diatoms, etc., also may be misleading, as there appear to be limits on the level to which the total number of species can be reduced without losing significant community information for comparative purposes [7,8]. Reducing community structure comparisons to less than one-third the species present may be the threshold.

DIVERSITY REDUCTION

Both above sets of problems potentially can be reduced. The conventional artificial substrate simulates hard surface habitat and is placed in shallow water. In the sea, the natural shallow water hard surface ecosystems are the algal forests. They are highly diverse ecosystems encompassing many phyla and many high-order taxonomic categories. Why should we deploy the artificial substrate concept into such diverse ecosystems? We can either seek naturally less diverse ecosystems for the experiments or modify the substrates to attract a lesser diversity of settlers.

Sediment tray substrates appear to be one of the answers. Boaden [9] and Renaud-Debsyer [10] have developed procedures, although Thorson [11] was using an inverted bottle float to collect sediments and infaunal larvae in the bottle neck (as well as glass-attaching forms on the bottle walls) in the arctic during the 1930s. Trays with well-sorted standardized sediment types (e.g., 0.5-mm sands, and 0.01-mm silts) held in shallow embayments at stated depths for stated times, etc., would induce settling by a relatively restricted array of infaunal species, for which there already exists a corps of benthic biologists trained in identification of the common species and in extraction of specimens of various size classes using conventions established over more than 80 years of quantitative infaunal studies [12].

The second development possibility is to utilize the low-diversity ecosystem between mid and high tides on rocky shores. The problem for standardization would be to dampen tidal oscillations to a standard range (say one meter), which could be achieved either computationally [13] or mechanically in some way.

CONCLUSIONS

The potential for standardizing artificial substrate protocols for multispecies bioassays within the sea lies in conventional hard substrates suitably strengthened for the turbulent nearshore environment, but also by use of sediment tray substrates. Realization of the potential is also dependent on adequate quality control of species identifications over a diversity of phyla, and on a standard decision-making procedure about whether to use single-number diversity indices based on low-cost species separation or high-cost species identifications and counts analyzed by species associational measures.

ACKNOWLEDGMENTS

I am much indebted to a number of scientific colleagues on whose work these comments are based. They include C. D. Levings, S. McKinell and S. Cross. In addition, I have been aware of the developments at industrial agencies through the cooperation of Rayonier Canada (BC) Ltd., Island Copper Mine and the Waste Management Branch of the BC Ministry of Environment.

REFERENCES

1. Adamson, N. E. "Technology of Ship-bottom Paints and Its Importance to Commercial and Naval Activities," *C & R Bull.* 10: 1-36 (1937).
2. Woods Hole Oceanographic Institution. *Marine Fouling and its Prevention,* Bureau of Ships, Navy Department (1951).
3. Clifford, H. T., and W. Stephenson. *An Introduction to Numerical Classification* (New York: Academic Press, Inc., 1975).
4. Green, R. H. *Sampling Design and Statistical Methods for Environmental Biologists* (New York: John Wiley & Sons, Inc.).
5. How, M. J. "The Application and Conduct of Ring Tests in Aquatic Toxicology," *Water Res.* 14(4): 293-296 (1980).
6. Ellis, D. V., and S. F. Cross. "A Protocol for Inter-laboratory Calibrations of Biological Species Identifications (Ring Tests)," *Water Res.* 15:1107-1108 (1981).
7. Moore, P. G. "Ecological Survey Strategy," *Mar. Poll. Bull.* 2: 37-39(1971).
8. Ellis, D. V., and K. E. Conlan. "Defining Species Associations for Pollution Assessment," *Can. Res. Tech. Section* 15-19(1979).
9. Boaden, P. J. S. "Colonization of Graded Sand by Interstitial Fauna," *Can. Biol. Mar.* 3:245-248 (1962).
10. Renaud-Debsyer, J. "Recherches écologiques sur la faune interstitielle des sables," *Vie Milieu, Suppl.* 15: 1-157(1963).
11. Thorson, G. "Technique and Future Work in Arctic Animal Ecology" *Meddr. Grnland* 144(4)(1946).
12. Holme, N. A., and A. D. McIntyre. *Methods For the Study of Marine Benthos,* IBP Handbook No. 16 (Oxford: Blackwell Scientific Publications, 1971).
13. Doty, M. S. "Critical Tide Levels that are Correlated with the Vertical Distribution of Marine Algae and Other Organisms along the Pacific Coast," *Ecology* 27 (4): 315-328(1946).

INDEX